KENKEN®

LIM-OPS, NO-OPS AND TWIST!

180 6 x 6 Puzzles That Make You Smarter

KenKen: Math & Logic Puzzles That Will Make You Smarter!
Series editor: Robert Fuhrer (*KenKen Puzzle, LLC, USA*)

ISSN: 2529-8003

KENKEN®
LIM-OPS, NO-OPS AND TWIST!

180 6 x 6 Puzzles That Make You Smarter

Created by
Tetsuya Miyamoto

Edited collection by
Robert Fuhrer
Founder, KenKen Puzzle, LLC

World Scientific

NEW JERSEY · LONDON · SINGAPORE · BEIJING · SHANGHAI · HONG KONG · TAIPEI · CHENNAI · TOKYO

Published by

World Scientific Publishing Co. Pte. Ltd.

5 Toh Tuck Link, Singapore 596224

USA office: 27 Warren Street, Suite 401-402, Hackensack, NJ 07601

UK office: 57 Shelton Street, Covent Garden, London WC2H 9HE

Library of Congress Cataloging-in-Publication Data
Names: Miyamoto, Tetsuya, 1959–   creator. | Fuhrer, Robert, editor.
Title: Kenken: lim-ops, no-ops and twist! : 180 6 × 6 puzzles that make you smarter / created by
    Tetsuya Miyamoto ; edited collection by Robert Fuhrer, founder, KenKen Puzzle Company.
Description: 1st edition. | New Jersey : World Scientific, [2020] |
    Series: Kenken: math & logic puzzles that will make you smarter!, 25298003 ; vol. 2
Identifiers: LCCN 2019052274 | ISBN 9789813236677 (hardcover) |
    ISBN 9789813235847 (paperback)
Subjects: LCSH: KenKen. | Logic puzzles. | Mathematical recreations.
Classification: LCC GV1493 .M5228 2020 | DDC 793.73--dc23
LC record available at https://lccn.loc.gov/2019052274

British Library Cataloguing-in-Publication Data
A catalogue record for this book is available from the British Library.

KenKen is a registered trademark of KenKen Puzzle, LLC. All rights reserved.

For any available supplementary material, please visit
https://www.worldscientific.com/worldscibooks/10.1142/10873#t=suppl

Printed in Singapore

CONTENTS

INTRODUCTION

Hello, puzzle enthusiast!

If you bought this book, there's a good chance you already know about KenKen. And if you already know about KenKen, then you likely know that solving KenKen puzzles will improve your calculating ability, logical thinking, problem solving, and patience.

But even if you're a KenKen fanatic, you may not know about all of the variations of KenKen. In this book, you'll be exposed to three of them: limited-operations puzzles, which limit the puzzle to only using *some* of the four operations, in this case, puzzles with only addition and subtraction but no multiplication or division; "no-ops" puzzles, which contain target numbers in the squares but no operations; and KenKen Twist, in which you'll use a different set of numbers to fill in the grid. If regular KenKen is "the puzzle that makes you smarter," then these variants will cause you to think in new ways and make you smarter still … though there may be a little frustration along the way, while you're trying to figure out the effects of these different rules. But, that's okay, isn't it? Change is good.

Limited-operations (or lim-ops) puzzles may sound like they'd be easier; after all, fewer operations mean fewer possibilities, right? Sadly, that's not the case. Multiplication and division often have fewer possibilities, because the factors help to eliminate some combinations. For instance, a [3−] cage in a 6×6 puzzle means the values could be {1, 4}, {2, 5}, or {3, 6}; whereas a [3×] cage would only have one possibility, {1, 3}. A [10+] cage with three cells in an L-shape? That would have six possibilities, namely {1, 3, 6}, {1, 4, 5}, {2, 2, 6}, {2, 3, 5}, {2, 4, 4}, or {3, 3, 4}. So, things can get ugly in a hurry.

With no-ops puzzles, things get even worse. A two-cell [3] cage in a 6×6 puzzle would have four possibilities for subtraction and multiplication, as shown above, but it would also have {1, 2} for addition.

(It would also have {1, 3} for division, but that's the same as multiplication.) And a three-cell [10] cage in an L-shape would have the same six combinations for addition as shown above, but also {1, 2, 5} for multiplication. My goodness! Hang on to your hat. You'll need to consider lots of combinations to solve these puzzles.

And then there's KenKen Twist. Pay attention with this one. Instead of using 1 through 6 for a 6×6 puzzle, you'll use a different candidate set each time. In one puzzle, you might use {1, 2, 3, 7, 8, 9}; in the next, {2, 4, 6, 7, 8, 9}. If you let down your guard, you may find yourself solving a puzzle with the wrong values!

The good news is that every puzzle in this book is a 6×6 puzzle. With all the other changes you'll be experiencing from these new rules, at least one thing will be consistent.

The even better news is that these variations will challenge you in new and fun ways. Sure, you may want to pull out your hair at first, but the long-term satisfaction you'll get from solving the puzzles in this book will greatly outweigh any momentary frustration.

Solutions can be found in the back of the book … but no cheating! Try solving each puzzle before checking your work.

The inventor of KenKen, Tetsuya Miyamoto, always says "The rival is not the person next to you. It's the YOU from yesterday. The more you challenge yourself, the more you will improve. Day by day, KenKen will help you grow!" To support Miyamoto-sensei's vision, there is a Solving Time box next to each puzzle so that you can track your progress and see how your skills improve. As you continue to solve KenKen puzzles, you will be able to increase your speed while maintaining accuracy.

Should you find that you need a break from all this craziness, visit our website at **www.kenkenpuzzle.com**, where you'll find an unlimited number of free, standard KenKen puzzles.

Thanks for taking a leap of faith and trying something new. Happy solving!

THE RULES OF STANDARD KENKEN

**Your goal is to fill in the whole grid
with numbers, making sure no number is
repeated in any row or column.**

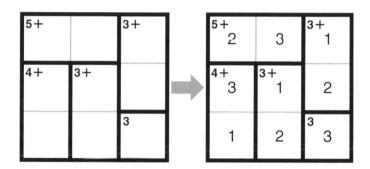

In a 3×3 puzzle, use the numbers 1–3.

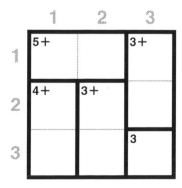

In a 4×4 puzzle, use the numbers 1–4.
In a 5×5, use the numbers 1–5, and so on.

5+		3+
4+	3+	
		3

The top left corner of each cage has a "target number" and math operation. The numbers you enter into a cage must combine (in any order) to produce the target number using the math operation noted (+, −, ×, or ÷)

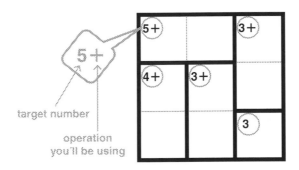

target number

operation
you'll be using

In this cage, the math operation to use is addition, **and the numbers must add up to 5. Since the cage has 2 squares, the only possibilities are 2 and 3, in either order (2+3 or 3+2=5).**

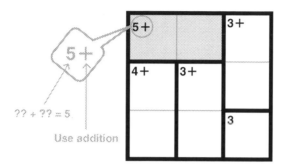

A cage with one square is a "Freebie"... just fill in the number you're given.

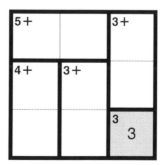

A number cannot be repeated within the
same row or column.

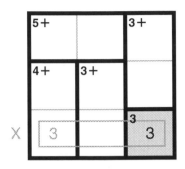

And that's it!

Solving a KenKen puzzle involves pure logic and mathematics.

No guesswork will ever be needed, and each puzzle has only one solution. So sharpen your pencil, sharpen your brain, and get started! In an instant, you'll know why Will Shortz, NPR Puzzle Master and The New York Times Puzzle editor calls KenKen "The most addictive puzzle since Sudoku!"

LIM-OPS PUZZLES
(Addition & Subtraction)

Step-by-Step Tutorial

Take it to the limit

In a typical KenKen puzzle, each cage contains a target number and an operation. In a lim-ops puzzle — short for "limited operations" — some of the operations won't be used. That is, the puzzle is limited to three, two, or even just one operation. In this volume, the lim-ops puzzles may contain only addition, only subtraction, or a mix of addition and subtraction. None will contain multiplication or division.

Hints for lim-ops puzzles

To solve lim-ops puzzles, you'll follow the same process as solving standard KenKen puzzles, but the following hints may prove helpful:

- Subtraction cages with a target number that is one less than the grid size can only have a single set of numbers as the answer. For instance, a [5−] cage in a 6×6 puzzle must be filled with {1, 6}.

- Addition cages with a target number that is one less than twice the grid size can only have a single set of numbers as the answer. For instance, a [11+] cage in a 6×6 puzzle must be filled with {5, 6}.

- Because every value is used exactly once in a row or column, the sum of the values is the same for every row and column in a puzzle. For a 6×6 puzzle, the sum of the values is always $1+2+3+4+5+6 = 21$. This can be hugely helpful. If a row contains, say, a two-cell [10+] cage and a three-cell [8+] cage, the sum of those two cages is $10+8 = 18$; since $21-18=3$, the remaining cell must be filled with a 3. Or if there is a two-cell [4−] cage in a column for which the other four cells have a sum of 13, then the cage must be filled with {2, 6}, because $21-13=8$ and $2+6=8$.

- Consider whether a value must be odd or even (mathematicians refer to this as the *parity* of a number). For instance, let's say a row in a 6×6 puzzle contains a three-cell [11+] cage and a two-cell [3−] cage. The possibilities for the [3−] cage are {1, 4}, {2, 5}, and {3, 6}. In every case, the sum of the possible values for the [3−] cage are odd, so the sum of the values in the [11+] cage and the [3−] cage must be even. Since the sum of the values in a row in a 6×6 puzzle is always 21 (odd), the remaining cell in that row must be odd. While this may not give you the value for the cell directly, it reduces the number of possibilities by half.

A lim-ops example

The lim-ops puzzle below contains both addition and subtraction.

6+	17+		9+		
			2−	9+	
3−	5+			8+	
	9+	9+			5−
10+		8+	4−		
			10+		

This puzzle has one "gimme." The [17+] cage must be filled with {5, 6, 6}, as follows:

6+	17+ 6	5	9+		
		6	2–	9+	
3–	5+			8+	
	9+	9+			5–
10+		8+	4–		
			10+		

Every row and column of a 6×6 puzzle uses each value 1 through 6 exactly once. Consequently, the sum of every row and every column in a 6×6 puzzle is $1+2+3+4+5+6=21$. This information is very helpful for all types of KenKen puzzles, but it is especially useful with lim-ops puzzles.

- Two cells in the top row of the puzzle contain {5, 6} in the [17+] cage, and there is a [9+] cage at the end of the row. The sum of those five cells is $5+6+9=20$, so the first cell in the row must be $21-20=1$.

- The two cells in the second row of the [6+] cage must contain {2, 3}, since a 1 was used in the top cell of that cage. Further, there is already a 6 in the third cell of the second row, and there's a [9+] cage at the end of the row. The sum of those five numbers is $2+3+6+9=20$, so the top cell of the [2–] cage must be $21-20=1$. And if the top cell in the [2–] cage is a 1, then the bottom cell of that cage must be a 3.

- The third row contains a [5+] cage and a [8+] cage, and the bottom cell of the [2–] cage is a 3. The sum of those five cells is $5+3+8=16$, so the first cell in the third row must be $21-16=5$. And if the top cell of the [3–] cage is a 5, then the bottom cell must be a 2.

- Since the [3–] cage contains a 2, the 2 in the [6+] cage must occur in the second column, and the 3 is then in the first column.

With just this little bit of work, the puzzle is more than a quarter complete:

6+	17+		9+		
1	6	5			
3	2	6	2– 1	9+	
3– 5	5+		3	8+	
2	9+	9+			5–
10+		8+	4–		
			10+		

As is typical with lim-ops puzzles, information from two (or more!) cages now needs to be used in tandem to make any progress.

- The [5+] cage cannot be filled with {2, 3}, because there is already a 3 in the third row. Therefore, the [5+] cage must be filled with {1, 4}. That means that the [8+] cage must be filled with {2, 6}. Since the [5–] cage in the last column must be filled with {1, 6}, the 6 in the [8+] cage must be in the left cell of that cage.

- The fourth row contains a 2 and a [9+] cage. Together, their sum is
  2+9=11, which means the other two cells must have a sum of 10.
  If the top cell of the [5−] cage were a 1, then the top cell of the [9+]
  cage would have to be a 9, which is impossible. Therefore, the 6 is in
  the top cell of the [5−] cage, and the top cell of the [9+] cage must
  be a 4.

Don't look now, but 16 of the 36 cells have been filled. Are we having
fun yet? You betcha!

6+ 1	17+ 6	5	9+		
3	2	6	2− 1	9+	
3− 5	5+		3	8+ 6	2
2	9+ 4	9+			5− 6
10+	5	8+	4−		1
			10+		

With nearly half of the puzzle complete, things start to move quickly.

- The [5+] cage must be filled with {1, 4}, and since there is already a 4
  in the second column, the order is dictated.

- The lower left cell of the three-cell [8+] cage must be 3, as it is now
  the only value not used in the second column.

- The two unfilled cells in the [8+] cage can either be {1, 4} or {2, 3}, but
  since a 4 already occurs in the third column, then it must be {2, 3}.
  Their order is determined because of the 3 in the lower left corner
  of the cage.

- The remaining unfilled cell in the third column (as part of the [9+] cage) must be a 1, as it is the only value not yet used.

- The other two cells in the [9+] cage must be {3, 5}, and the 3 in the [2–] cage above dictates their order.

The puzzle is now two-thirds complete.

6+ 1	17+ 6	5	9+		
3	2	6	2– 1	9+	
3– 5	5+ 1	4	3	8+ 6	2
2	9+ 4	9+ 1	5	3	5– 6
10+	5	8+ 3	4–		1
	3	2	10+		

One last piece of logic is needed to crest the hill.

- The bottom row contains {2, 3} in the [8+] cage, and there is also a [10+] cage. The sum of these five numbers is $2 + 3 + 10 = 15$, so the first cell in the bottom row must be $21 - 15 = 6$. The other number in the two-cell [10+] cage must be 4.

The hard work is now done, and the remainder of the puzzle becomes a formality.

- The only numbers missing from the fifth row are {2, 6}, and the 6 used in the [8+] cage in the third row dictates the order of {2, 6} in the [4–] cage.

- The three-cell [10+] cage in the bottom row must be filled with {1, 4, 5}, and the 1's and 5 in the cages above are enough to dictate their order.

- The two-cell [9+] cage in the second row must be filled with {4, 5}, and the three-cell [9+] cage in the first row must be filled with {2, 3, 4}, and their order is dictated by the other numbers already in those columns.

And there you have it, *c'est fini.*

6+	17+		9+		
1	6	5	2	4	3
3	2	6	2− 1	9+ 5	4
3− 5	5+ 1	4	3	8+ 6	2
2	9+ 4	9+ 1	5	3	5− 6
10+ 4	5	8+ 3	4− 6	2	1
6	3	2	10+ 4	1	5

As this puzzle shows, lim-ops puzzles can be deceptive. With only two operations, they appear to be less complex than other KenKens, but don't let their simplicity fool you. Addition and subtraction cages often have more possible combinations than multiplication or division cages, so more information is needed to eliminate the incorrect candidates.

No pain, no gain. Lim-ops puzzles may take you longer than other types of KenKens, but you'll also feel a great sense of satisfaction (and relief!) when you complete one, because of the effort that is required.

6×6 Lim-Ops Puzzles (Addition & Subtraction)

Let's start with 6×6 addition and subtraction puzzles. The goal is to fill each cell using only the numbers 1, 2, 3, 4, 5, and 6 without repeating a number in any row or column.

1

6+	4−		11+	3−	
	11+			7+	11+
		13+			
4−				6+	9+
9+	5−				
	2−		5−		

SOLVING TIME

Tip: *Remember, a "freebie" is always a great place to start. In Puzzle 2, fill in the "4" near the bottom left corner first, and you're on your way!*

2

11+		5−		11+	
	2−		5−		
5−	2−		7+		8+
	3−	4−	5+	9+	
4					
3+		1−		11+	

SOLVING TIME

3

9+		10+		5−	
5−			10+		
3+	5+	2−		16+	7+
2−	5+		11+		6+
	8+		4		

SOLVING TIME

4

4−		8+		5+	
11+		8+	3+	17+	
5−					1−
	9+		2−		
8+			5−		1−
	7+		6+		

Tip: Take a few seconds to look over the full puzzle before you begin. It's a great way to quickly pick out some of the easier-to-solve cages.

SOLVING TIME

5

8+		3	3−		17+
	3+		2−		
4	9+			4−	
3−	5	5−		7+	13+
	14+	1−			
			1		

SOLVING TIME

6

12+	7+		3+		18+
		5−			
7+			8+	4−	
18+	5−				4+
		2−	3−	11+	
3					

SOLVING TIME

7

3+		10+	11+	11+	
6+				5	
	5	9+		5−	
11+		1		7+	
10+		9+	1		1−
6			3+		

SOLVING TIME

8

9+		3−	4+	11+	
5−					8+
	6+		11+		
8+	10+		13+		3+
		8+			
3−		3−		5+	

9

5+		11+		3−	
10+		2−		3+	
8+	3−		2−		14+
		3−	5−		
12+				11+	
	3+		9+		

SOLVING TIME

10

3	10+	5−		13+	
		18+			6+
8+	3			4+	
	2−		5		2−
15+		3+		4	
	3+		8+		6

SOLVING TIME

11

3−	3−	11+	3−		3
			6+		10+
3−		13+			
11+	8+			7+	
	3		1−	5−	
	5−			9+	

SOLVING TIME

12

1−		4−		15+	
2	7+		5−		
5−	11+		3	8+	5+
	13+	2	2−		
10+					5+
		1−		3	

Tip: *As with baking, ping pong, and playing the bassoon, when it comes to KenKen, practice makes perfect. The more puzzles you do, the faster your brain will process the next one! Keep going ... you'll see exactly what we mean.*

13

5−	7+		19+		2
	7+		11+		
8+	5+			2	
	5−		2−	3−	5+
13+		3			
4		1−		4−	

SOLVING TIME

SOLVING TIME

14

14

19+					2
5+		1−		1	19+
5−	8+		14+		
		7+			
10+		5−		4−	
3		3−			

SOLVING TIME

15

2	8+		15+		
5−			14+		
7+	4−	4−		3−	
			5−		14+
16+					
	15+				

SOLVING TIME

15

16

3−	22+				2−
		2−	4−	11+	
7+					
5−	9+	3+		8+	4
			5		4−
1−		10+			

17

5+	9+		12+		5+
	15+	2−			
		2−	5+	7+	
11+				5−	
5−		11+		12+	5+
	5+				

18

10+	7+		11+	8+	
				1	1−
3+		14+			
2−	17+				
	3	7+		10+	
11+			3−		

19

2−	11+		7+		13+
	3+		2−		
1−		9+	6+	6+	
8+					
14+			11+		3−
	5−		1−		

20

2−		4+		17+	
4+	7+		22+		2−
				13+	
11+					2−
	8+		3−		
4		5−		2−	

21

5	3+	11+			13+
3+		9+	6+		
	3−			1−	5−
10+		4+	14+		
	12+				
		11+			

22

3−		6	19+		
3	2−	3+		2−	
5−			6+		5
	1−				5+
2−	2−	8+	10+	5	
				5−	

SOLVING TIME

23

4−	2	5−		1−	
	1−	8+	15+		
3			8+		
13+			8+	1−	1
11+					2−
5−		10+			

SOLVING TIME

24

13+		4−		5+	
	7+			1−	
2−	3−		4−		1−
	11+		3+		
3+		12+			9+
8+		2−			

SOLVING TIME

25

1−		12+	5+	12+	6
1−					
	4	6+	15+	4−	
22+	9+			12+	
					3+
			2−		

SOLVING TIME

26

2−	6+	5−		3−	3−
		13+	10+		
5−					5+
	12+		3−		
2−			15+		
6+				6	

SOLVING TIME

27

4	13+	10+		4−	
			16+	5−	
5−				2	15+
		8+			
4−		5−		15+	
7+					3

SOLVING TIME

28

5+		6	21+		
6+			10+		
3	1−		3+		
17+			3−	10+	8+
12+	5				
			2		

SOLVING TIME

29

10+		4−		3−	
3		10+	11+		15+
11+			1		
	12+		12+		
				10+	2
5−		1−			

SOLVING TIME

30

5−	12+				1−
	12+		21+		
1−		3+	6+		
	17+			18+	
8+				3	
		7+			

SOLVING TIME

31

1−		1−		5−	
9+		23+			1
	4	9+	3−		
1−				10+	
15+					10+
2−		4+			

SOLVING TIME

32

8+	7+	7+		11+	
			5−		8+
		2−	6+	17+	
1−	6				
	5+		10+		
2−		5−			

33

9+	15+			18+	
	3+		3−		
	19+	3−	2−	1	
4−				16+	
		5−			6+
4					

34

17+			9+		
	6+	1−		1−	
		4−	8+	9+	5+
23+					
	11+			5−	
		7+			

SOLVING TIME

35

5+	8+	1−		9+	
			24+		
3−	5−	5			
		1−	2−	2−	
9+	9+			3−	1−
		5−			

SOLVING TIME

36

15+			14+		
	2	7+	6+	1−	
	16+			2−	
				5−	
2−	11+	8+	10+	8+	
					1

37

22+					3
	10+	6+	15+	2	23+
5−					
		2	7+		
15+		6+			
3		5+			

38

1−	5+	5−		4	28+
			2−		
4+					
1−		17+			
4−	10+		6+		
		1−			

SOLVING TIME

39

5−		17+			
1−		1−		4−	
9+	2−	1−			5+
		8+		2−	
8+			6		11+
	10+				

SOLVING TIME

40

7+			15+	3−	
16+	4−	1−		1−	
				10+	
		12+	1−	1−	
4−					
1−			1−		1

41

16+			1−	7+	
13+		5−		5	
	6+		10+		
			2−		6
	11+			11+	
5	14+				

42

17+				2−	
2−		6	2−		1−
2−		2−	9+	2	
4−	2			2−	
	5−		2	1−	
3−		12+			

SOLVING TIME

43

12+		3	1−	9+	
	11+				13+
6+	2−		8+		
			10+		
	5−	16+			1−
5					

SOLVING TIME

44

21+			5−		6+
	5		4+		
6+		15+			
	11+	2	25+		
10+			6+		
		2−			

SOLVING TIME

45

19+	3−		7+	5−	2
					15+
9+			5		
	10+		12+		1−
17+					
		11+			

SOLVING TIME

46

2−	20+		9+		5−
		3−		2−	
		10+	8+		5+
11+				1−	
3−		1−			1−
	4+		7+		

SOLVING TIME

47

9+		4	5−	11+	
15+	2−				19+
	9+			13+	9+
1−	1−				
	3	1−			

SOLVING TIME

48

20+	6	1−		4−	
					2−
19+	3+		6+	13+	
					2−
11+		2−			
3		9+		5−	

49

15+	4−		3	7+	14+
3+		15+			
12+	10+			7+	
	13+		12+		
		1	4−		

50

17+				2−	
6	11+		10+		
1−		13+	5		
			3−		
2−	14+	3	5−	8+	

51

12+			7+	17+	
6+					
7+	10+		14+	1	9+
	1				
16+		5+	8+	1−	
6					

52

16+				8+	
5+		3	1−		
6+		15+			
17+		8+		1−	
	8+			8+	
		10+			

53

3	27+		2−		10+
11+	1				
		4+			6+
2−			10+		
	7+			1	11+
15+					

54

6+		4−		9+	
5−		7+	17+	3−	
1−					
	12+		10+		6
3−	2−			8+	
			5		

55

15+			3+	14+	1
17+	2−				
		4+	2−	13+	
7+					
		4−	2−		
4			14+		

56

9+		3+		12+	
4	19+				
3+		2−	5+	25+	6
					5+
12+					
		2−		5+	

SOLVING TIME

57

5	22+			7+	
5−		8+			5+
			9+		
2−	3−		10+	4−	2−
	9+	5+			
				10+	

SOLVING TIME

36

58

9+		6+		8+	
	10+			4−	15+
4		11+	5+		
4−	5−			15+	
					6+
1−		7+			

59

18+			14+	5+	
	5			14+	
		8+			5−
2−			10+		
10+	3−			11+	
	11+				

60

9+			4+	11+	
3−	11+	19+			5−
				14+	
2−			1−		7+
	15+				
5−					2

SOLVING TIME

Great job!

You have conquered the art of solving Lim-Ops KenKens!

It's time to move on to No-Ops puzzles.

NO-OPS PUZZLES
(All Operations)

Step-by-Step Tutorial

Cancel the surgery ... no operations!

In a traditional KenKen puzzle, each cage contains a target number and an operation. The values to be placed in that cage must combine, in any order, to yield the target number using the given operation. For instance, the numbers in a [10×] cage must yield 10 when multiplied together.

With no-ops puzzles, each cage contains a target number but *no operation*. It means that *any* operation can be used to yield the target number. For instance, in the no-ops puzzle below, the numbers that fill the [10] cage must yield 10 when addition, subtraction, multiplication, or division is applied. The challenge is, you don't know which operation to use, so most cages in a no-ops puzzle will have far more possibilities than the cages in a standard KenKen puzzle. If it were a [10+] cage, the values would have to be {4, 6}. If it were a [10×] cage, the values would have to be {2, 5}. But since it's a [10] cage in a no-ops puzzle, it could be filled with either combination.

Hints for no-ops puzzles

To solve no-ops puzzles, you'll follow the same process as solving standard KenKen puzzles, but the following hints may prove helpful:

- Look for big numbers. In a 6×6 no-ops puzzle, a two-cell cage with a target number greater than 12, a three-cell cage with a target number greater than 18, and, in general, an *n*-cell cage with a target number greater than 6*n* means that the operation must be multiplication.

- Look for prime numbers larger than the grid size. For instance, in the puzzle above, the [11] and [17] cages imply addition, because it's not possible to get a product of either 11 or 17 with multiplication. (If the prime number is less than the grid size, however, such as 2, 3, or 5 in the case of a 6×6 puzzle, every operation is a possibility.)

A no-ops example

It's not clear where to start with the 6×6 no-ops puzzle below. That's often the case; it may take some exploration of what you know before you'll be able to fill in any cell.

6		17		240	
	2			11	
8	11				
	90			12	
15			4		6
		10			

Most of the cages in the puzzle above have multiple possibilities. As already mentioned above, the [10] cage has two possibilities: {4, 6} with addition or {2, 5} with multiplication. The [15] cage is a bit more challenging, because it could be either addition or multiplication, and values can be repeated in the cage because of its L-shape. Consequently, there are three possibilities for the [15] cage: {1, 3, 5}, {4, 5, 6}, {3, 6, 6}.

But the [17] cage at the top must be addition — no three values can multiply to 17 — and the only possibility is {5, 6, 6}. Because it is an L-shaped cage, the placement of each value is uniquely determined. This is kind of a big deal.

- The [17] cage requires two 6's, so the [10] cage in the bottom row cannot contain a 6. Consequently, it must be filled with {2, 5}, and because there is a 5 in the [17] cage, the placement of the 5 in the bottom row is determined.

- The [11] cage in the fifth column must be {5, 6}, and since there is already a 6 in the second row, the order of the values is determined.

Together, these insights form a very good start.

The [240] cage must be multiplication, because no other operation could generate a target number of that magnitude.

- The possibilities are {2, 4, 5, 6} or {3, 4, 4, 5}, but a 6 has already been placed in each of the first three rows. Hence, it must be {3, 4, 4, 5}. Because the 5 must be used in the third row, and because the set contains two 4's, the placement of all four values is uniquely determined.

- The [12] cage can either be {3, 4} or {2, 6}, but since a 4 has now been used in both the fifth and sixth columns, it must be {2, 6}, and their order is dictated by the values above.

We're getting somewhere! In fact, we're more than one-third done already.

6		**17** 6	5	**240** 4	3
	2		6	**11** 5	4
8	**11**			6	5
	90			**12** 2	6
15			**4**		**6**
		10 5	2		

Things are really starting to fall into place!

- The [8] cage can be either {2, 6} with addition or {2, 4} with multiplication. But {2, 6} is already used in the [12] cage in the fourth row, so the bottom cell in the [8] cage can't be 2 or 6. Hence, it must be 4, with the 2 above.

- The last two cells in the sixth column must be {1, 2}, which means the other cell in the [6] cage must be a 3; the operation can be either addition or multiplication. Further, the 2 in the [10] cage dictates the position of the {1, 2}.

- Similar logic applies to the [6] cage in the upper left, with {1, 2} completing the top row. Hence, the other cell in that cage must be a 3, and the order of the {1, 2} in the top row is dictated by the {2, 4} in the [8] cage.

There are still a lot of unfilled cells in the center of the puzzle, but our hard work has resulted in good progress. Twenty-one of the cells are now filled.

6		17		240	
1	2	6	5	4	3
3	2		6	11	4
				5	
8	11				
2				6	5
	90			12	
4				2	6
15			4	6	
					2
		10			
		5	2	3	1

Much of the rest of the puzzle can be filled by identifying the missing values in a row or column.

- The [15] cage must have {5, 6} in the first column and {4, 6} in the bottom row, so the 6 must occupy the lower left cell. The placements of the 4 and 5 are then determined.

- The [2] cage in the second row must contain {1, 2}, with the placement dictated by the 2 in the [6] cage above.

- After the 1 is placed in the [2] cage, the three unfilled cells remaining in the second column must contain {3, 5, 6} in some order, and the 6's and 5 already placed in the third and fourth rows imply their placements.

With the exception of the [4] cage in the fifth row, every cage is now at least partially completed.

6		17		240	
1	2	6	5	4	3
3	**2** 1	2	6	**11** 5	4
8 2	**11** 3			6	5
4	**90** 5			**12** 2	6
15 5	6		**4**		**6** 2
6	4	**10** 5	2	3	1

Just seven cells left to go! A final bit of sleuthing will allow us to complete the puzzle.

- The [4] cage can be filled with either {1, 3} by addition or {1, 4} by multiplication. The important realization is that there must be a 1 somewhere in the cage.

- With a {5, 6} already in the [90] cage, the other two cells must be {1, 3}. Because a 1 must occur in the [4] cage, the bottom right cell of the [90] cage must be 3, with the 1 above it.

- Consequently, the last unfilled cell in the third column must be a 4, and the two remaining cells in the [11] cage must be {1, 3}.

- With {1, 3} filling the two cells in the fourth column of the [11] cage, the position of the {1, 4} in the [4] cage is now determined.

Finally, we're ready to place the last of the values and complete the puzzle.

6		17		240	
1	2	6	5	4	3
	2			11	
3	1	2	6	5	4
8	11				
2	3	4	1	6	5
	90			12	
4	5	1	3	2	6
15			4		6
5	6	3	4	1	2
		10			
6	4	5	2	3	1

And that does it! That wasn't so bad, was it?

Of course, the solution that appears above was written *after* the puzzle was solved, which makes it look like the solution was effortless. Far from it! To find that elegant solution required stops and starts. You'll experience the same thing when you solve some no-ops puzzles on your own, but be persistent! If you continue to explore the cages and consider possibilities, you'll eventually find something that works, and unlocking one door will likely unlock others.

6×6 No-Ops Puzzles (All Operations)

61

160			18	5	
7					3
	5		12	2	
3		9		1	
	270			24	10

62

4	3		12		15
12	32		1350		
				10	24
15		5		9	
11		2			1

63

3	15	11		32	
		13	30		
60					
	2		54		
	2			13	
		1			6

64

5	80		4		15
		3	11		
4	3			108	
16		90	10		32
		1			

65

1	2		54		
	15	10			
1		720			
	3				3
4		3	1		
		8		10	

SOLVING TIME

66

50		480	54	7	
					72
			2		
72		5	11		
5				40	
	6				5

SOLVING TIME

67

11		4	6		
9	11		3		16
		3			
7		3		11	
2		6	21	2	

SOLVING TIME

68

2	120			6	
	32	60		1	
			3	8	120
30	60				
			108		
				2	

SOLVING TIME

4	14			11	18
450	3				
	4	5		30	
		12			2
3				80	
18			2		

SOLVING TIME

5	120	6		4	2
				5	
15		144			
8		4		3	
2		6	150		2
	3				

SOLVING TIME

71

72			5	2	40
1	9			180	
6					
60	5	4	60	2	
		7			1
	4		6		

72

96			45		11
	2	5			
	6	25		2	
10			16		24
		12			
11			3		

5	24		12	108	
5	2				
			2		4
15	23		11		
			5		16

100	12			36	
		5		3	
3	30		1		
	4			22	
108		7			
	3		5		

SOLVING TIME

SOLVING TIME

75

11		2	15		
5	80			54	
					2
1	19			2	5
	8		21		

76

11		90	2		2
9			2		
6	3			2	
		2	16	2	
	13				10
6					

77

1	1		6	6	
	7	400			
		7	3		5
3	120				2
4			90		
		2		2	

78

240	2	54	10		5
				2	24
	3		4	11	
		5	6		
120		96		2	
3					

79

11	1		12	3	24
	5				
	8	6		96	
3					50
		2	30		
24				3	

SOLVING TIME

80

40	11		6		
			13		
12		11		2	
90		16	90	24	
		1		1	

SOLVING TIME

81

180	10		2		3
	40		1	3	
		120			40
	6			5	
5		3	2		
4	3		1		

SOLVING TIME

82

180		2		3	
	15	5	1		2
2			3	7	1
		150			
1	5				180
7		3			

SOLVING TIME

83

288			1	10	
	6	4	12		
			100	2	
3				11	5
30	8	7			
			2		4

SOLVING TIME

84

1	16		72		
160	3				1
		14		6	
14			7	2	
		48		45	

SOLVING TIME

85

9	12			24	
		90		11	
12				40	5
9	5	4	2		8
		1			

86

8		1	1800		
	7			3	20
2	14				
3			72	7	5
200	5				

10		3	240		
6			15	90	
1	24				
		2			72
1		10	5		
30					

SOLVING TIME

72	13			3	
		1	5		96
10					
6		24	1	1	
				30	
9		5		2	

SOLVING TIME

54		14	30	9	2
	2				
			5		1
9		10	6		
5				24	
4		2			3

SOLVING TIME

5	60			17	2
	2		36		
2	1				8
	11		1		
19			3		15
		7			

SOLVING TIME

91

450		2	19		
				30	
5		4	45		8
12					
	11			2	
		15			3

92

14		15		2	2
	21		14		
6					
	30		6	13	5
7					
			72		

93

2	3	90	18		
				1920	
2	40				4
	6				
4	2	120			
			2		3

SOLVING TIME

94

2	5		6		20
	15			13	
3	6	24			
		15		72	
5			8		2
48					

SOLVING TIME

95

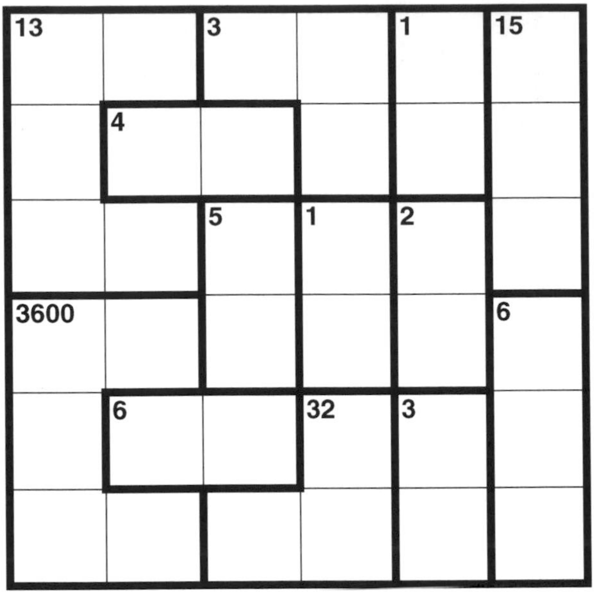
13
3
1
15
4
5
1
2
3600
6
6
32
3

SOLVING TIME

96

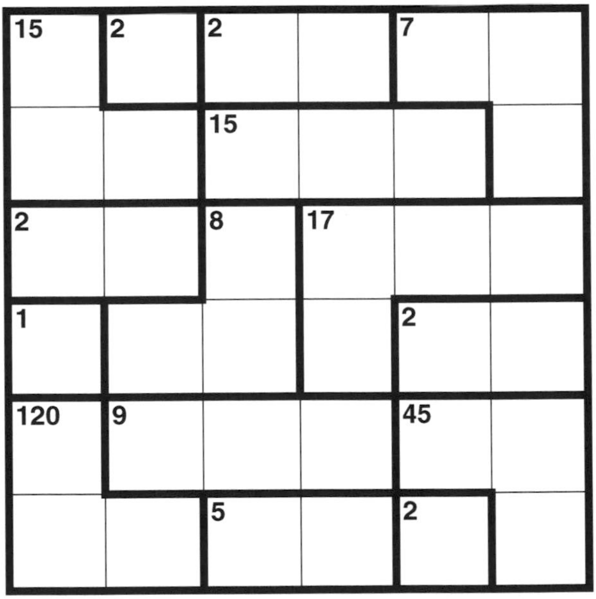
15
2
2
7
15
2
8
17
1
2
120
9
45
5
2

SOLVING TIME

97

144				12	
3		3	2	60	
3	21				
			15	5	
3					72
11					

98

2	864		30	24	
				150	5
		12			
	45		7		
		4		3	
15				2	

99

100		5		13	2
	2				8
14		90			
			24	3	
		20			1
3			2		

100

2	60		15		
		36	10		
				1	
20	144			8	
		4		5	
			3		3

101

30		15	2		
90					12
		1	2		
9			11		
	4	48			8
2		1			

SOLVING TIME

102

5		3		72	
	22				30
3		1	7		
		11	2	1	
24					
	13				6

SOLVING TIME

103

6	40				2
12	15	2		90	
		1	54		
				3	2
	2		12		
10					4

104

6			10		4
3	300			11	
2					3
	2880				
30		3	7		1

105

11			15	50	
	9				5
3				2	
4320	12		6		1
				4	
				2	

SOLVING TIME

106

26			3		2
	4		2		
	2		6	1	2
	24				
2		20		14	2
	3				

SOLVING TIME

107

100		2	2	6	
					5
2	120			4	
		1350			4
9			12		
4	4				

108

480		5	1		15
10				10	
		2	4		
				3	
108		2	60		
			6		2

109

180		13	24		
			6	4	1
	3	60			
		9		8	
	4				18
4					

110

48			9		
3	1	3	20	144	1
	18				
11		15			2
		1	4		

SOLVING TIME

SOLVING TIME

111

112

113

114

115

576				6	
	6	14		5	
2				1440	
	1		5		
6	20		3		
7		3		4	

SOLVING TIME

116

90		3	2		3
	2		8		
2		17			9000
2			2		
	10				14

SOLVING TIME

117

3	25		5		11
			5		
1			48	11	
15					8
	2	15		5	
		2			

SOLVING TIME

118

20		4	2		4
	5		12		
9		5		24	
	16		10		
1		16			150

SOLVING TIME

119

1	72		9		1
		8			
10	3	2		864	2
	6				
		40		14	
6					

120

50			2		7
1		1	3		
	6			30	24
3		3			
	480			3	
6					

Great job!

You have conquered the No-Ops KenKens in this book!

Ready for a new challenge?

It's time to move on to Twist puzzles.

TWIST PUZZLES
(All Operations)

Step-by-Step Tutorial

C'mon, baby, let's do the KenKen Twist

In a typical KenKen puzzle, the numbers 1 through n are used in an $n \times n$ grid. That is, the numbers 1 through 3 are used in a 3×3 puzzle, the numbers 1 through 4 are used in a 4×4 puzzle, and so on. But in KenKen Twist, you'll use a different set of numbers. Those numbers will be identified above the grid. For example, in the 4×4 KenKen puzzle below, instead of using 1, 2, 3, and 4, the numbers to be used are 2, 4, 6, and 8.

2,4,6,8

48×		8+	
12+		4−	
	8		20+
2÷			

Otherwise, the regular rules of KenKen still apply.

Hints for KenKen Twist

To solve KenKen Twist puzzles, you'll follow the same process as solving standard KenKen puzzles, but the following hints may prove helpful:

- If you don't know which number goes in a cell, note the possibilities! Use what you know to eliminate as many possibilities as you can. (This is the first and most important rule for solving any KenKen puzzle: standard, no-ops, twist, or otherwise.)

- Note the factors of the numbers in the candidate set. For instance, for the set {2, 4, 6, 8}, the number 6 is the only one with a factor of 3. (The other three numbers are all powers of 2.) This is important; you will know immediately if a multiplication cage contains a 6 or not, based on whether the target number is a multiple of 3. If the target number is a multiple of 9, then the cage would contain two 6's.

- Look for special numbers in the candidate set, such as prime numbers. If 5 and 7 appear in the candidate set, they'll provide a lot of information, especially regarding multiplication cages.

- If one number in the candidate set is significantly larger than the others, it's a strong possibility to be involved in large sums. For instance, if a [25+] cage has just four cells, and the candidate set is {1, 3, 4, 9}, it's a sure bet that at least one 9 will appear in the cage.

- Consider the collection of numbers in the candidate set. For example, the candidate set {2, 3, 5, 7} contains only prime numbers, so the multiplication and subtraction cages will be unique; addition cages will contain at least one 2 if there are an odd number of cells and the target number is even; and, division is impossible. Similarly, the candidate set {1, 2, 4, 8} contains only powers of 2, which means there will be many possibilities for multiplication and division, but subtraction will have unique target numbers; and the candidate set {2, 4, 6, 8}, which was used for the puzzle above, contains only even numbers.

A KenKen Twist example

Instead of the numbers 1 through 6, the 6×6 KenKen Twist puzzle below uses the candidate set {1, 2, 3, 7, 8, 9}. Fill in the cages with the values 1, 2, 3, 7, 8, or 9 so that the given operation yields the target number. As always, you may only use each number once in each column and row.

1,2,3,7,8,9

18+		21+	4÷		7×
			4×		
5—				27+	
8—	14×	5—			
		15+		1728×	3
2÷					

Before you begin, consider the candidate set. You can think of it as bimodal, since three of the numbers (1, 2, 3) are small, and three of the numbers (7, 8, 9) are big. There is a gap between these two subsets. That may be important when solving the puzzle, since 4, 5, and 6 won't be used.

In addition, take note of those cages in the puzzle that must be filled with unique numbers:

• Of course, the [3] cage will be filled with 3.

• The [2÷] cage in the last row must be filled with {1, 2}, because none of the three large numbers (7, 8, 9) in the candidate set yield a quotient of 2 when divided by any of the three small numbers (1, 2, 3).

• The [4÷] cage can only be filled with {2, 8}.

• The three-cell [4×] cage can only be filled with {1, 2, 2}.

• The [7×] cage must be filled with {1, 1, 7}.

• The [14×] cage must be filled with {2, 7}.

- The [8–] cage must be filled with {1, 9}. (This is equivalent to a [5–] cage appearing in a standard 6×6 KenKen puzzle.)

- Finally, the [1728×] cage is equal to 12×12×12, which is equal to $2^6 \times 3^3$. This is an important point: to get six factors of 2, two 8's will be needed; and then to get three factors of 3, a 3 and a 9 will be needed. Consequently, this cage will be filled with {3, 8, 8, 9}, which means that an 8 will appear in the top cell of the cage.

Based on these observations alone, we're able to fill 11 cells in the puzzle.

1,2,3,7,8,9

18+		21+	4÷ 8	2	7× 1
			4× 2	1	7
5–		2	1	27+	
8–	14×	5–			
		15+		1728× 3 8	3
2÷					8

That's an auspicious beginning!

There are now some additional observations that can be made:

- In the sixth column, the two remaining cells must be {2, 9}, since the other four numbers have been used. Then, the other two cells in the [27+] cage must be {7, 9} to get the required target number.

- Since {7, 9} will be used in the fifth column for the [27+] cage, then the remaining {3, 9} in the [1728×] cage cannot have the 9 in the fifth column.

- Since {1, 2} will be used in the [2÷] cage, and since {3, 8, 9} are used in the sixth row as part of the [1728×] cage, then the remaining cell in the bottom row must be 7. Consequently, {1, 7} must be used in the other two cells of the [15+] cage, and their order is dictated.

Adding in these additional pieces gets us more than halfway home, and we're well on the way to a full solution.

1,2,3,7,8,9

18+		21+	4÷ 8	2	7× 1
			4× 2	1	7
5−		2	1	27+ 7	9
8−	14×	5−		9	2
		15+ 1	7	1728× 8	3 3
2÷		7	9	3	8

Because of the pieces that are already filled in, the order of the values in the [8−], [14×], and [2÷] cages are dictated and can be filled in:

- The [8−] cage must be filled with {1, 9}, and the 9 cannot be in the top cell.

- The [14×] cage must be filled with {2, 7}, and the 7 cannot be in the bottom cell.

- The [5–] cage in the fourth row must be filled with {3, 8}, because a 2 has already been used in that row. Further, the 8 in the [4÷] cage of the first row dictates their order.

- The [2÷] cage must be filled with {1, 2}, and the {2, 7} in the [14×] cage dictates their order.

That's a lot of information, and filling in those values means the puzzle is three-fourths complete.

1,2,3,7,8,9

18+		21+	4÷ 8	2	7× 1
			4× 2	1	7
5–			2	1	27+ 7	9
8– 1	14× 7	5– 8	3	9	2
9	2	15+ 1	7	1728× 8	3 3
2÷ 2	1	7	9	3	8

Finally, the following can be noted:

- The numbers {3, 9} must fill the remaining cells in the third column, so the [21+] cage must be filled with {3, 9, 9}.

- That leaves {3, 7, 8} to fill the [18+] cage.

- The only candidates remaining for the [5–] cage in the third row are {3, 8}.

The placement of these values is determined by the other numbers in each row and column, so this solves the puzzle.

1,2,3,7,8,9

18+		21+	4÷		7×
7	3	9	8	2	1
8	9	3	4× 2	1	7
5− 3	8	2	1	27+ 7	9
8− 1	14× 7	5− 8	3	9	2
9	2	15+ 1	7	1728× 8	3 3
2÷ 2	1	7	9	3	8

Do not be fooled! The step-by-step solution above makes it appear that the solution was easily found, but it wasn't! Before a concise, lucid solution can be obtained, many false starts and mistakes may occur. Do not be discouraged if you have trouble solving KenKen puzzles or if your solution process isn't as smooth as the solution above. That's normal! Remember, you're a rock star for even trying these puzzles, and the point is to have fun, improve your problem-solving ability, and develop patience. Don't stress. Enjoy!

6×6 Twist Puzzles (All Operations)

121

1,3,5,6,7,8

22+		**3−**		**320×**	
		3÷			**1−**
5−	**6**		**280×**	**24×**	
	120×				**5**
		30×		**7**	**18×**
	15+				

SOLVING TIME

122

1,3,5,6,7,8

3−	**144×**	**8+**		**5**	**3÷**
		42×	**2÷**		
			1−		**30×**
735×		**3−**	**48×**		
6×				**3÷**	**1−**
		2−			

SOLVING TIME

86

123

1,3,5,6,7,8

5−	42×		5−		280×
	3÷		13+		
450×		21×		27+	
5−			7		108×
280×					

124

1,3,5,6,7,8

5	5−	40×	2−		1−
2÷			13+		
	2−		56×	9+	
16+	3÷			11+	
	56×	42×	2÷		75×

125

2,4,6,7,8,9

7−	3−	512×			3−
			15+	9+	
2−	9408×				21+
	11+				
224×				54×	
		5−		3÷	

126

2,4,6,7,8,9

5−	1−	42×		15+	2÷
		2÷			
2−	5−	5−		4÷	1−
		5−	3−		
4−				336×	
23+					7

127

2,4,6,7,8,9

48×		2÷		2−	
11+	1728×		512×		3÷
		2−			
108×		42×			288×
	5−	19+		42×	
		4			

128

2,4,6,7,8,9

112×			6804×		4
17+		48×	13+		
2−	108×				
			2÷		21+
20+		2−		1−	
		7−			

129

3,4,5,7,8,9

3240×	15+		567×	1−	
				3−	5−
20+					
		22+		3÷	35×
480×					
	84×			1−	

SOLVING TIME

130

3,4,5,7,8,9

18+		14+		12×	
9+		1−	15+	17+	
	2520×			120×	7
3÷					13+
	2÷		525×		
7		13+			

SOLVING TIME

90

131

3,4,5,7,8,9

19+	3÷		15+		40×
		108×			
	2÷	3−		4−	23+
17+		3−			
	7	12+			
315×			5−		

132

3,4,5,7,8,9

35×		3÷		20+	
2−	16+				3÷
	17+	4	22+		
2÷			9		3−
	648×	120×			
		1−		35×	

133

2,3,5,6,8,9

18×	3—		27+		6
	2—		1—		
1—	3—			3	
	19+		3÷	45×	2—
14+					
8		11+		7—	

134

2,3,5,6,8,9

3÷	11+	48×	5—	2700×	
					4÷
6—	3÷	1—			
			28+		54×
30×	4÷	4—	3÷		

135

2,3,5,6,8,9

810×				720×	
28+				3	
	54×	5	2÷	36+	
13+		19+			36×
6	3−				

SOLVING TIME

136

2,3,5,6,8,9

360×	17+			11+	1−
		810×			
			48×		6−
3÷	8+		3−		
	24×		2	7+	3÷
9	19+				

SOLVING TIME

137

1,2,4,6,7,9

1944×			2÷		7
392×			13+	1296×	
	2	5−			
			18+		
10+	8−			5−	3+
		13+			

SOLVING TIME

138

1,2,4,6,7,9

9	36×	13+		14×	12+
2÷		5−			
	112×	8−		3888×	
5−					3÷
	20+			36×	

SOLVING TIME

139   *1,2,4,6,7,9*

5−	2÷	28×		19+	
		20+	14×		
4	29+				1−
5292×		1	16+		
				9+	
			1		

SOLVING TIME

140   *1,2,4,6,7,9*

7−	3÷	4	378×	20+	
				7	
5−			2	1728×	
	18+				9
252×				2÷	
		14×		7+	

SOLVING TIME

141

4,5,6,7,8,9

32+				180×	
4	315×	3−			2÷
			56×		
1−	11+		3−	63×	
	5	8640×		48×	
3−					

SOLVING TIME

142

4,5,6,7,8,9

22680×	15+	30×	15+		120×
			2÷		
			3−		
30+			2−		2520×
	2−	36×	7		
			4−		

SOLVING TIME

143

4,5,6,7,8,9

12+		168×	21+	1−	
2÷				10080×	
120×	15+			5−	
		405×	18+		
432×			140×		

SOLVING TIME

144

4,5,6,7,8,9

5−	4	56×		1−	
	2−		29+	2−	
				1−	
15+		24+			6
3−			63×		2÷
1−		45×		4	

SOLVING TIME

145

1,2,3,5,6,7

35×		16+		12+	5−
3÷					
3−			5−		15+
5−	2÷		35×		
	6	42×			
7	17+				

146

1,2,3,5,6,7

4−		4−	5−	105×	
360×				5−	
	9+				6−
		35×	2÷		
4−			2÷		3÷
6	8+		8+		

147

1,2,3,5,6,7

84×			15×		
19+		4−		10×	
	6+		2÷		42×
		10×			
17+		5−		18+	
	6+				

SOLVING TIME

148

1,2,3,5,6,7

11+		14×		36×	
10+		3÷			13+
		15×	30×		
1−			350×	3÷	
3−		18+		1−	

SOLVING TIME

99

149

2,4,5,6,8,9

180×	17+			9	1152×
		17+			
	2−		5−		1−
4÷			3−		
	12960×		6	15+	
6					

SOLVING TIME

150

2,4,5,6,8,9

5−		2÷	8	34+	
4÷	5		180×		11+
	4320×				
24×		23+			1−
			9+		
		6		4−	

SOLVING TIME

151

2,4,5,6,8,9

2÷		432×			29+
11+					
18×	1−			10+	
	30×		1280×	4÷	
15+		8			
2−		1−		9	

SOLVING TIME

152

2,4,5,6,8,9

288×		2÷		1−	
3−		5	72×		16+
	19+				
48×		19+			72×
	22+		2	32×	
9		1−			

SOLVING TIME

101

1,3,4,7,8,9

7×	216×		1−	15+	21+
18144×			33+	1	
	36×				
		196×		9×	
	8		4÷		

SOLVING TIME

1,3,4,7,8,9

3×		24+			4
168×		96×	5−	54432×	
					3÷
	18+		6−		
1008×					
		19+			

SOLVING TIME

155

1,3,4,7,8,9

21×	4÷	1−		16+	21+
		4−			
	24×		4÷		
24+		4−		21×	
	37+				
		11+			1

156

1,3,4,7,8,9

27×	168×			4÷	
		20+			7−
1−		19+	216×		
8+				16+	
	18+			84×	
1−		8+			

157

1,2,3,5,6,7

2100×					3
	11+	1	16+		27+
8+		2÷			
		3−		3	
630×		6+			
		4−			

158

1,2,3,5,6,7

12×		8+		5	630×
	17+				
2−				3÷	
	882×		6	2÷	
2		1−	6×		15+

159

1,2,3,5,6,7

11+	14×	15×	1−		3÷
			14×	6+	
5−	2−	5−			4−
			2−	1−	
3÷	5−	5−			2−
			1−		

160

1,2,3,5,6,7

9+	6×	3÷	10×		2−
			7+	2−	
6×		15+			6
1			84×		
28+	1−			1−	
				2−	

2,4,5,6,7,9

192×		3−		34020×	
210×					252×
		22+			
2÷				1−	
405×			7	12+	
	12+		2−		

SOLVING TIME

2,4,5,6,7,9

5−	3−	10080×			2
			7−		270×
21+		7			
		36×	2−	3150×	
1−	3÷				28×

SOLVING TIME

163

2,4,5,6,7,9

63×	5−		1−		3−
	22+		33+		
4−		150×			2÷
20×	3÷	567×		2−	12+

SOLVING TIME

164

2,4,5,6,7,9

60×	2÷		15+		29+
	15+		5		
	3−		5−		
36×		21+		1−	15+
	210×				
7			2÷		

SOLVING TIME

107

165

1,3,4,6,8,9

26+			162×		
	18+			8−	
		5−	7+	2−	
7776×				3÷	
	2÷		4÷		8
			21+		

SOLVING TIME

166

1,3,4,6,8,9

4÷	21+			12×	2−
		72×	23+		
					27×
144×			288×		
1−	3888×	1			
					4

SOLVING TIME

167

1,3,4,6,8,9

2−		4÷		3÷	
10+		34992×	5−	2÷	
	3−			10+	
72×					
	72×		5−		2−
13+			7−		

SOLVING TIME

168

1,3,4,6,8,9

12×		13+	144×		
17+			8−		6
3−	10+			18+	
	29+				
7−	48×		2÷		1−
	8−				

SOLVING TIME

169

3,4,5,6,8,9

35+			1280×		
	2−		2−		
		192×			3÷
69120×	22+				
	15×		36×		16+
			3		

170

3,4,5,6,8,9

720×	2−		14+		1−
		26+			
	16200×				72×
1296×					
		2÷		150×	
11+		9	24×		

171

3,4,5,6,8,9

31104×		4−	2÷		21+	
			160×	5	
					3−
40×	12+		12960×		
	4			15+	
9	2−				

SOLVING TIME

172

3,4,5,6,8,9

90×		6	1−		13+
	3	192×	3−		
5			21+	1−	33+
2÷	36×				
		30×			
17+				5	

SOLVING TIME

111

173

2,3,5,7,8,9

5	18+		280×		3÷
96×				3÷	
		41+	80×		7
3÷					4÷
		5−	6−	14+	

SOLVING TIME

174

2,3,5,7,8,9

5−		9450×			7
			13+		36+
16+		147×	9		
4÷					
	5	216×			6×
24+					

SOLVING TIME

175

2,3,5,7,8,9

4÷	88200×			30×	8
		3	16+		
3÷		18+			36+
			7		
210×					
		4÷		2−	

176

2,3,5,7,8,9

2−	54×		20+		4÷
		4−	19+		
56448×					2−
9		4÷		2835×	
		50×	63×		

177

1,2,4,5,7,9

27+				90×	
	5−			18+	
7−	4−	3−			5−
		2÷			
3−	5−		450×		7×

178

1,2,4,5,7,9

98×		4−		20+	
	2	4÷	3−	4−	
8−	18+				
		9+	2	8+	2520×
20×					
		9			

179

1,2,4,5,7,9

28×	16+			2÷	315×
		7−	5−		
3−					
18×			1960×		
26+				5−	
	7+				

SOLVING TIME

180

1,2,4,5,7,9

9	2÷		33+		
14×				1−	
504×		20×			8−
			2520×		
21+				4÷	
10+					

SOLVING TIME

115

Ken-gratulations!

You have conquered all the puzzles in this book!

Play more KenKen in other books and solve unlimited puzzles online!

SOLUTIONS

1

6+ 3	**4−** 6	2	**11+** 5	**3−** 4	1
1	**11+** 3	4	6	**7+** 2	**11+** 5
2	4	**13+** 1	3	5	6
4− 6	2	5	4	**6+** 1	**9+** 3
9+ 5	**5−** 1	6	2	3	4
4	**2−** 5	3	**5−** 1	6	2

2

11+ 5	3	**5−** 6	1	**11+** 4	2
3	**2−** 4	2	**5−** 6	1	5
5− 1	**2−** 6	4	**7+** 5	2	**8+** 3
6	**3−** 5	**4−** 1	**5+** 2	**9+** 3	4
4 4	2	5	3	6	1
3+ 2	1	**1−** 3	4	**11+** 5	6

3

9+ 4	5	**10+** 3	2	**5−** 1	6
5− 6	1	5	**10+** 3	4	2
3+ 2	**5+** 3	**2−** 6	1	**16+** 5	**7+** 4
1	2	4	5	6	3
2− 3	**5+** 4	1	**11+** 6	2	**6+** 5
5	**8+** 6	2	**4** 4	3	1

4

4− 2	6	**8+** 5	3	**5+** 4	1
11+ 4	2	**8+** 3	**3+** 1	**17+** 5	6
5− 1	5	4	2	6	**1−** 3
6	**9+** 4	1	**2−** 5	3	2
8+ 5	3	2	**5−** 6	1	**1−** 4
3	**7+** 1	6	**6+** 4	2	5

5

8+ 1	4	**3** 3	**3−** 5	2	**17+** 6
3	**3+** 1	2	**2−** 4	6	5
4 4	**9+** 3	6	2	**4−** 5	1
3− 2	**5** 5	**5−** 1	6	**7+** 3	**13+** 4
5	**14+** 6	**1−** 4	3	1	2
6	2	5	**1** 1	4	3

6

12+ 5	**7+** 3	4	**3+** 2	1	**18+** 6
2	5	**5−** 1	6	3	4
7+ 1	4	2	**8+** 3	**4−** 6	5
18+ 4	**5−** 1	6	5	2	**4+** 3
6	2	**2−** 3	**3−** 4	**11+** 5	1
3 3	6	5	1	4	2

7

3+ 1	2	**10+** 6	**11+** 5	**11+** 4	3
6+ 2	1	3	6	**5** 5	4
4	**5** 5	**9+** 2	3	**5−** 6	1
11+ 5	6	**1** 1	4	**7+** 3	2
10+ 3	4	**9+** 5	**1** 1	2	**1−** 6
6 6	3	4	**3+** 2	1	5

8

9+ 4	1	**3−** 5	3	**4+** 2	**11+** 6
5− 6	4	2	1	3	**8+** 5
1	**6+** 2	4	**11+** 5	6	3
8+ 3	**10+** 6	1	**13+** 4	5	**3+** 2
5	3	**8+** 6	2	4	1
3− 2	5	**3−** 3	6	**5+** 1	4

9

5+ 2	3	11+ 6	5	3- 1	4
10+ 6	4	2- 5	3	3+ 2	1
8+ 1	3- 6	3	2- 2	4	14+ 5
5	2	3- 4	5- 1	6	3
12+ 4	5	1	6	11+ 3	2
3	3+ 1	2	9+ 4	5	6

10

3 3	10+ 4	5- 6	1	13+ 2	5
1	5	18+ 3	4	6	6+ 2
8+ 2	3 3	5	6	4+ 1	4
6	2- 2	4	5 5	3	2- 1
15+ 5	6	3+ 1	2	4 4	3
4	3+ 1	2	8+ 3	5	6 6

11

3- 6	3- 2	11+ 5	3- 4	1	3 3
3	5	6	6+ 1	2	10+ 4
3- 1	4	13+ 2	5	3	6
11+ 4	8+ 1	3	6	7+ 5	2
5	3 3	4	1- 2	5- 6	1
2	5- 6	1	3	9+ 4	5

12

1- 3	2	4- 1	5	15+ 4	6
2 2	7+ 4	3	5- 1	6	5
5- 1	11+ 5	6	3 3	8+ 2	5+ 4
6	13+ 3	2 2	2- 4	5	1
10+ 5	6	4	2	1	5+ 3
4	1	1- 5	6	3 3	2

13

5- 6	7+ 3	4	19+ 1	5	2 2
1	2	5	11+ 6	4	3
8+ 3	5+ 4	1	5	2 2	6
5	5- 1	6	2- 2	3- 3	5+ 4
13+ 2	5	3 3	4	6	1
4 4	6	1- 2	3	4- 1	5

14

19+ 5	6	1	3	4	2 2
5+ 2	3	1- 4	5	1 1	19+ 6
5- 1	8+ 4	2	14+ 6	3	5
6	2	7+ 3	4	5	1
10+ 4	5	5- 6	1	4- 2	3
3 3	1	3- 5	2	6	4

15

2 2	8+ 3	1	15+ 4	6	5
5- 1	6	4	14+ 5	3	2
7+ 4	4- 5	4- 6	3	3- 2	1
3	1	2	5- 6	5	14+ 4
16+ 5	2	3	1	4	6
6	15+ 4	5	2	1	3

16

3- 2	22+ 1	5	4	6	2- 3
5	6	2- 3	4- 2	11+ 4	1
7+ 3	4	1	6	2	5
5- 6	9+ 3	3+ 2	1	8+ 5	4 4
1	2	4	5 5	3	4- 6
1- 4	5	10+ 6	3	1	2

17

2 [5+]	3 [9+]	6	1 [12+]	5	4 [5+]
3	5 [15+]	2 [2−]	4	6	1
4	6	1 [2−]	3 [5+]	2 [7+]	5
5 [11+]	4	3	2	1 [5−]	6
1 [5−]	2	5 [11+]	6	4 [12+]	3 [5+]
6	1 [5+]	4	5	3	2

18

1 [10+]	4 [7+]	3	5 [11+]	2 [8+]	6
3	2	4	6	1 [1]	5 [1−]
2 [3+]	1	5 [14+]	3	6	4
4 [2−]	5 [17+]	6	2	3	1
6	3 [3]	1 [7+]	4	5 [10+]	2
5 [11+]	6	2	1 [3−]	4	3

19

1 [2−]	5 [11+]	6	3 [7+]	4	2 [13+]
3	1 [3+]	2	4 [2−]	6	5
4 [1−]	3	5 [9+]	1 [6+]	2 [6+]	6
6 [8+]	2	4	5	1	3
2 [14+]	4	3	6 [11+]	5	1 [3−]
5	6 [5−]	1	2 [1−]	3	4

20

2 [2−]	4	1 [4+]	3	5 [17+]	6
1 [4+]	3 [7+]	4	5 [22+]	6	2 [2−]
3	1	5	6	2 [13+]	4
6 [11+]	5	3	2	4	1 [2−]
5	6 [8+]	2	4 [3−]	1	3
4 [4]	2	6 [5−]	1	3 [2−]	5

21

5 [5]	1 [3+]	2 [11+]	6	3	4 [13+]
1 [3+]	2	4 [9+]	5 [6+]	6	3
2	3 [3−]	5	1	4 [1−]	6 [5−]
4 [10+]	6	3 [4+]	2 [14+]	5	1
6	4 [12+]	1	3	2	5
3	5	6 [11+]	4	1	2

22

5 [3−]	2	6 [6]	1 [19+]	3	4
3 [3]	4 [2−]	1 [3+]	5	2 [2−]	6
1 [5−]	6	2	3 [6+]	4	5 [5]
6	5 [1−]	4	2	1	3 [5+]
4 [2−]	1 [2−]	3 [8+]	6 [10+]	5 [5]	2
2	3	5	4	6 [5−]	1

23

5 [4−]	2 [2]	1 [5−]	6	4	3 [1−]
1	3 [1−]	2 [8+]	4 [15+]	6	5
3 [3]	4	6	1 [8+]	5	2
4 [13+]	6	3	5 [8+]	2 [1−]	1 [1]
2 [11+]	5	4	3	1	6 [2−]
6 [5−]	1	5 [10+]	2	3	4

24

6 [13+]	4	1 [4−]	5	3 [5+]	2
3	1 [7+]	2	4	6 [1−]	5
2 [2−]	6 [3−]	3	1 [4−]	5	4 [1−]
4	5 [11+]	6	2 [3+]	1	3
1 [3+]	2	5 [12+]	3	4	6 [9+]
5 [8+]	3	4 [2−]	6	2	1

1 ⁽¹⁻⁾	**2**	**5** ⁽¹²⁺⁾	**3** ⁽⁵⁺⁾	**4** ⁽¹²⁺⁾	**6** ⁽⁶⁾
4 ⁽¹⁻⁾	**1**	**6**	**2**	**5**	**3**
3	**4** ⁽⁴⁾	**2** ⁽⁶⁺⁾	**6** ⁽¹⁵⁺⁾	**1** ⁽⁴⁻⁾	**5**
2 ⁽²²⁺⁾	**3** ⁽⁹⁺⁾	**1**	**5**	**6** ⁽¹²⁺⁾	**4**
5	**6**	**3**	**4**	**2**	**1** ⁽³⁺⁾
6	**5**	**4** ⁽²⁻⁾	**1**	**3**	**2**

5 ⁽²⁻⁾	**2** ⁽⁶⁺⁾	**1** ⁽⁵⁻⁾	**6**	**4** ⁽³⁻⁾	**3** ⁽³⁻⁾
3	**4**	**2** ⁽¹³⁺⁾	**5** ⁽¹⁰⁺⁾	**1**	**6**
1 ⁽⁵⁻⁾	**5**	**6**	**3**	**2**	**4** ⁽⁵⁺⁾
6	**3** ⁽¹²⁺⁾	**4**	**2** ⁽³⁻⁾	**5**	**1**
4 ⁽²⁻⁾	**6**	**5**	**1** ⁽¹⁵⁺⁾	**3**	**2**
2 ⁽⁶⁺⁾	**1**	**3**	**4**	**6** ⁽⁶⁾	**5**

4 ⁽⁴⁾	**6** ⁽¹³⁺⁾	**2** ⁽¹⁰⁺⁾	**3**	**1** ⁽⁴⁻⁾	**5**
3	**4**	**5**	**2** ⁽¹⁶⁺⁾	**6** ⁽⁵⁻⁾	**1**
1 ⁽⁵⁻⁾	**3**	**4**	**5**	**2** ⁽²⁾	**6** ⁽¹⁵⁺⁾
6	**2**	**3** ⁽⁸⁺⁾	**1**	**5**	**4**
5 ⁽⁴⁻⁾	**1**	**6** ⁽⁵⁻⁾	**4**	**3** ⁽¹⁵⁺⁾	**2**
2 ⁽⁷⁺⁾	**5**	**1**	**6**	**4**	**3** ⁽³⁾

4 ⁽⁵⁺⁾	**1**	**6** ⁽⁶⁾	**5** ⁽²¹⁺⁾	**3**	**2**
2 ⁽⁶⁺⁾	**3**	**1**	**4** ⁽¹⁰⁺⁾	**6**	**5**
3 ⁽³⁾	**4** ⁽¹⁻⁾	**5**	**1** ⁽³⁺⁾	**2**	**6**
6 ⁽¹⁷⁺⁾	**2**	**4**	**3** ⁽³⁻⁾	**5** ⁽¹⁰⁺⁾	**1** ⁽⁸⁺⁾
1 ⁽¹²⁺⁾	**5** ⁽⁵⁾	**2**	**6**	**4**	**3**
5	**6**	**3**	**2** ⁽²⁾	**1**	**4**

5 ⁽¹⁰⁺⁾	**3**	**6** ⁽⁴⁻⁾	**2**	**4** ⁽³⁻⁾	**1**
3 ⁽³⁾	**2**	**5** ⁽¹⁰⁺⁾	**4** ⁽¹¹⁺⁾	**1**	**6** ⁽¹⁵⁺⁾
2 ⁽¹¹⁺⁾	**4**	**3**	**1** ⁽¹⁾	**6**	**5**
1	**5** ⁽¹²⁺⁾	**2**	**6** ⁽¹²⁺⁾	**3**	**4**
4	**6**	**1**	**3**	**5** ⁽¹⁰⁺⁾	**2** ⁽²⁾
6 ⁽⁵⁻⁾	**1**	**4** ⁽¹⁻⁾	**5**	**2**	**3**

6 ⁽⁵⁻⁾	**2** ⁽¹²⁺⁾	**4**	**5**	**1**	**3** ⁽¹⁻⁾
1	**3** ⁽¹²⁺⁾	**5**	**6** ⁽²¹⁺⁾	**4**	**2**
3 ⁽¹⁻⁾	**4**	**1** ⁽³⁺⁾	**2** ⁽⁶⁺⁾	**6**	**5**
4	**6** ⁽¹⁷⁺⁾	**2**	**3**	**5** ⁽¹⁸⁺⁾	**1**
2 ⁽⁸⁺⁾	**5**	**6**	**1**	**3** ⁽³⁾	**4**
5	**1**	**3** ⁽⁷⁺⁾	**4**	**2**	**6**

3 ⁽¹⁻⁾	**2**	**5** ⁽¹⁻⁾	**4**	**1** ⁽⁵⁻⁾	**6**
5 ⁽⁹⁺⁾	**3**	**2** ⁽²³⁺⁾	**6**	**4**	**1** ⁽¹⁾
1	**4** ⁽⁴⁾	**3** ⁽⁹⁺⁾	**2** ⁽³⁻⁾	**6**	**5**
2 ⁽¹⁻⁾	**1**	**6**	**5**	**3** ⁽¹⁰⁺⁾	**4**
6 ⁽¹⁵⁺⁾	**5**	**4**	**1**	**2**	**3** ⁽¹⁰⁺⁾
4 ⁽²⁻⁾	**6**	**1** ⁽⁴⁺⁾	**3**	**5**	**2**

1 ⁽⁸⁺⁾	**2** ⁽⁷⁺⁾	**4** ⁽⁷⁺⁾	**3**	**5** ⁽¹¹⁺⁾	**6**
4	**3**	**2**	**6** ⁽⁵⁻⁾	**1**	**5** ⁽⁸⁺⁾
2	**1**	**5** ⁽²⁻⁾	**4** ⁽⁶⁺⁾	**6** ⁽¹⁷⁺⁾	**3**
5 ⁽¹⁻⁾	**6** ⁽⁶⁾	**3**	**2**	**4**	**1**
6	**4** ⁽⁵⁺⁾	**1**	**5** ⁽¹⁰⁺⁾	**3**	**2**
3 ⁽²⁻⁾	**5**	**6** ⁽⁵⁻⁾	**1**	**2**	**4**

33

[9+]1	[15+]4	5	6	[18+]3	2
5	[3+]1	2	[3−]3	6	4
3	[19+]5	[3−]4	[2−]2	[1]1	6
[4−]6	2	1	4	[16+]5	3
2	3	[5−]6	1	4	[6+]5
[4]4	6	3	5	2	1

34

[17+]2	4	6	[9+]5	3	1
1	[6+]2	4	3	[1−]6	5
4	3	[4−]1	[8+]6	[9+]5	[5+]2
[23+]6	1	5	2	4	3
3	[11+]5	2	4	[5−]1	6
5	6	3	[7+]1	2	4

35

[5+]1	[8+]3	[1−]4	5	[9+]6	2
4	2	3	[24+]6	1	5
[3−]2	[5−]1	[5]5	3	4	6
5	6	[1−]2	[2−]4	[2−]3	1
[9+]6	[9+]4	1	2	[3−]5	[1−]3
3	5	[5−]6	1	2	4

36

[15+]4	1	2	[14+]5	6	3
6	[2]2	[7+]1	[6+]3	[1−]5	4
2	[16+]4	6	1	[2−]3	5
5	3	4	2	[5−]1	6
[2−]1	[11+]5	[8+]3	[10+]6	[8+]4	2
3	6	5	4	2	[1]1

37

[22+]2	1	6	5	4	[3]3
4	[10+]3	[6+]1	[15+]6	[2]2	[23+]5
[5−]1	2	5	3	6	4
6	5	[2]2	[7+]4	3	1
[15+]5	4	[6+]3	2	1	6
[3]3	6	[5+]4	1	5	2

38

[1−]3	[5+]2	[5−]6	1	[4]4	[28+]5
4	1	2	[2−]3	5	6
[4+]1	3	4	5	6	2
[1−]5	6	[17+]3	4	2	1
[4−]6	[10+]5	1	[6+]2	3	4
2	4	[1−]5	6	1	3

39

[5−]1	6	[17+]5	3	4	2
[1−]4	5	[1−]1	2	[4−]6	3
[9+]3	[2−]4	[1−]6	5	2	[5+]1
6	2	[8+]3	1	[2−]5	4
[8+]2	1	4	[6]6	3	[11+]5
5	[10+]3	2	4	1	6

40

[7+]1	2	4	[15+]5	[3−]3	6
[16+]3	[4−]1	[1−]2	4	[1−]6	5
4	5	3	6	[10+]1	2
6	3	[12+]1	[1−]2	[1−]5	4
[4−]2	6	5	1	4	3
[1−]5	4	6	[1−]3	2	[1]1

41

16+ 3	6	5	**1−** 4	**7+** 2	1
13+ 6	2	**5−** 1	3	**5** 5	4
2	**6+** 3	6	**10+** 1	4	5
4	1	2	**2−** 5	3	**6** 6
1	**11+** 5	4	2	**11+** 6	3
5 5	**14+** 4	3	6	1	2

42

17+ 4	5	2	6	**2−** 3	1
2− 2	4	**6** 6	**2−** 3	1	**1−** 5
2− 3	1	**2−** 5	**9+** 4	**2** 2	6
4− 1	**2** 2	3	5	**2−** 6	4
5	**5−** 6	1	**2** 2	**1−** 4	3
3− 6	3	**12+** 4	1	5	2

43

12+ 4	2	**3** 3	**1−** 6	**9+** 5	1
6	**11+** 4	1	5	3	**13+** 2
6+ 2	**2−** 3	4	**8+** 1	6	5
3	5	2	**10+** 4	1	6
1	**5−** 6	**16+** 5	2	4	**1−** 3
5 5	1	6	3	2	4

44

21+ 5	2	4	**5−** 6	1	**6+** 3
4	**5** 5	6	**4+** 1	3	2
6+ 2	3	**15+** 5	4	6	1
1	**11+** 4	**2** 2	**25+** 3	5	6
10+ 3	6	1	**6+** 2	4	5
6	1	**2−** 3	5	2	4

45

19+ 5	**3−** 1	4	**7+** 3	**5−** 6	**2** 2
2	3	6	4	1	**15+** 5
9+ 1	2	3	**5** 5	4	6
6	**10+** 5	2	**12+** 1	3	**1−** 4
17+ 4	6	1	2	5	3
3	4	**11+** 5	6	2	1

46

2− 4	**20+** 2	5	**9+** 3	6	**5−** 1
2	5	**3−** 1	4	**2−** 3	6
5	3	**10+** 4	**8+** 6	1	**5+** 2
11+ 1	4	6	2	**1−** 5	3
3− 3	6	**1−** 2	1	4	**1−** 5
6	**4+** 1	3	**7+** 5	2	4

47

9+ 3	6	**4** 4	**5−** 1	**11+** 2	5
15+ 5	**2−** 1	3	6	4	**19+** 2
4	5	2	3	6	1
6	**9+** 2	1	4	**13+** 5	**9+** 3
1− 1	**1−** 4	5	2	3	6
2	**3** 3	**1−** 6	5	1	4

48

20+ 4	**6** 6	**1−** 3	2	**4−** 5	1
1	3	4	6	2	**2−** 5
19+ 6	**3+** 1	2	**6+** 5	**13+** 4	3
2	5	6	1	3	**2−** 4
11+ 5	4	**2−** 1	3	6	2
3 3	2	**9+** 5	4	**5−** 1	6

49

15+ 2	4− 1	5	3: 3	7+ 4	14+ 6
6	4	3	2	1	5
3+ 1	2	15+ 4	5	6	3
12+ 4	10+ 3	6	1	7+ 5	2
5	13+ 6	2	12+ 4	3	1
3	5	1: 1	4− 6	2	4

50

17+ 5	3	1	2	2− 6	4
1	6: 6	11+ 2	3	10+ 4	5
3	1− 2	6	13+ 4	5: 5	1
2	1	4	5	3− 3	6
2− 4	14+ 5	3: 3	5− 6	8+ 1	2
6	4	5	1	2	3

51

12+ 1	3	5	7+ 6	17+ 4	2
6+ 2	4	3	1	6	5
7+ 4	10+ 2	6	14+ 5	1: 1	9+ 3
3	1: 1	2	4	5	6
16+ 5	6	5+ 1	8+ 2	1− 3	4
6: 6	5	4	3	2	1

52

16+ 5	3	6	2	8+ 1	4
5+ 1	4	3: 3	1− 5	6	2
6+ 3	2	15+ 5	6	4	1
17+ 2	1	8+ 4	3	1− 5	6
4	8+ 6	2	1	8+ 3	5
6	5	10+ 1	4	2	3

53

3: 3	27+ 5	2	2− 6	4	10+ 1
11+ 2	1: 1	5	4	3	6
1	2	4+ 3	5	6	6+ 4
2− 4	6	1	10+ 3	5	2
6	7+ 3	4	2	1: 1	11+ 5
15+ 5	4	6	1	2	3

54

6+ 2	4	4− 5	1	9+ 6	3
5− 1	6	7+ 4	17+ 3	3− 5	2
1− 4	2	1	6	3	5
5	12+ 1	3	10+ 2	4	6: 6
3− 3	2− 5	6	4	8+ 2	1
6	3	2	5: 5	1	4

55

15+ 5	6	4	3+ 2	14+ 3	1: 1
17+ 6	2− 3	5	1	4	2
2	4	4+ 3	2− 6	13+ 1	5
7+ 3	5	1	4	2	6
1	2	4− 6	2− 3	5	4
4: 4	1	2	14+ 5	6	3

56

9+ 3	6	3+ 1	2	12+ 4	5
4: 4	19+ 5	6	3	1	2
3+ 1	2	2− 3	5+ 4	25+ 5	6: 6
2	3	5	1	6	5+ 4
12+ 6	4	2	5	3	1
5	1	2− 4	6	5+ 2	3

57

5 **5**	22+ **6**	**4**	**1**	7+ **3**	**2**
5− **6**	**4**	8+ **3**	**5**	**2**	5+ **1**
1	**2**	**5**	9+ **3**	**6**	**4**
2− **2**	3− **3**	**6**	10+ **4**	4− **1**	2− **5**
4	9+ **1**	5+ **2**	**6**	**5**	**3**
3	**5**	**1**	**2**	10+ **4**	**6**

58

9+ **3**	**4**	6+ **1**	**5**	8+ **6**	**2**
2	10+ **3**	**4**	**1**	4− **5**	15+ **6**
4 **4**	**2**	11+ **6**	5+ **3**	**1**	**5**
4− **1**	5− **6**	**5**	**2**	15+ **3**	**4**
5	**1**	**2**	**6**	**4**	6+ **3**
1− **6**	**5**	7+ **3**	**4**	**2**	**1**

59

18+ **5**	**1**	**4**	14+ **6**	5+ **3**	**2**
2	5 **5**	**1**	**3**	14+ **6**	**4**
3	**2**	8+ **6**	**5**	**4**	5− **1**
2− **1**	**3**	**2**	10+ **4**	**5**	**6**
10+ **4**	3− **6**	**3**	**1**	11+ **2**	**5**
6	11+ **4**	**5**	**2**	**1**	**3**

60

9+ **4**	**3**	**2**	4+ **1**	11+ **6**	**5**
3− **5**	11+ **4**	19+ **3**	**2**	**1**	5− **6**
2	**5**	**4**	**6**	14+ **3**	**1**
2− **1**	**2**	**6**	1− **4**	**5**	7+ **3**
3	15+ **6**	**1**	**5**	**2**	**4**
5− **6**	**1**	**5**	**3**	**4**	2 **2**

61

160 **4**	**2**	**5**	18 **6**	5 **3**	**1**
7 **2**	**4**	**6**	**5**	**1**	3 **3**
5	5 **6**	**1**	12 **3**	2 **2**	**4**
3 **3**	**1**	9 **2**	**4**	1 **5**	**6**
1	270 **3**	**4**	**2**	24 **6**	10 **5**
6	**5**	**3**	**1**	**4**	**2**

62

4 **4**	3 **1**	**3**	12 **2**	**6**	15 **5**
12 **6**	32 **2**	**4**	1350 **5**	**1**	**3**
2	**4**	**1**	**3**	10 **5**	24 **6**
1	**3**	**5**	**6**	**2**	**4**
15 **3**	**5**	5 **6**	**1**	9 **4**	**2**
11 **5**	**6**	2 **2**	**4**	**3**	1 **1**

63

3 **2**	15 **3**	11 **6**	**5**	32 **1**	**4**
6	**5**	13 **3**	30 **1**	**4**	**2**
60 **1**	**6**	**4**	**2**	**5**	**3**
5	2 **4**	**2**	54 **6**	**3**	**1**
4	2 **2**	**1**	**3**	13 **6**	**5**
3	**1**	1 **5**	**4**	**2**	6 **6**

64

5 **6**	80 **5**	**4**	4 **2**	**1**	15 **3**
1	**4**	3 **3**	11 **6**	**2**	**5**
4 **4**	3 **2**	**1**	**5**	108 **3**	**6**
16 **3**	**1**	90 **5**	10 **4**	**6**	32 **2**
2	**3**	**6**	**1**	**5**	**4**
5	**6**	1 **2**	**3**	**4**	**1**

5 ⁽¹⁾	**4** ⁽²⁾	**2**	**1** ⁽⁵⁴⁾	**3**	**6**
6	**5** ⁽¹⁵⁾	**1** ⁽¹⁰⁾	**2**	**4**	**3**
4 ⁽¹⁾	**3**	**6** ⁽⁷²⁰⁾	**5**	**2**	**1**
3	**2** ⁽³⁾	**4**	**6**	**1**	**5** ⁽³⁾
1 ⁽⁴⁾	**6**	**3** ⁽³⁾	**4** ⁽¹⁾	**5**	**2**
2	**1**	**5** ⁽⁸⁾	**3**	**6** ⁽¹⁰⁾	**4**

2 ⁽⁵⁰⁾	**5**	**4** ⁽⁴⁸⁰⁾	**3** ⁽⁵⁴⁾	**6** ⁽⁷⁾	**1**
5	**1**	**2**	**6**	**3**	**4** ⁽⁷²⁾
3	**4**	**5**	**2** ⁽²⁾	**1**	**6**
4 ⁽⁷²⁾	**6**	**1** ⁽⁵⁾	**5** ⁽¹¹⁾	**2**	**3**
1 ⁽⁵⁾	**3**	**6**	**4**	**5** ⁽⁴⁰⁾	**2**
6	**2** ⁽⁶⁾	**3**	**1**	**4**	**5** ⁽⁵⁾

6 ⁽¹¹⁾	**5**	**4** ⁽⁴⁾	**2** ⁽⁶⁾	**3**	**1**
3 ⁽⁹⁾	**6** ⁽¹¹⁾	**5**	**4** ⁽³⁾	**1**	**2** ⁽¹⁶⁾
5	**1**	**6** ⁽³⁾	**3**	**2**	**4**
4 ⁽⁷⁾	**3**	**2** ⁽³⁾	**1**	**6** ⁽¹¹⁾	**5**
2 ⁽²⁾	**4**	**1** ⁽⁶⁾	**6** ⁽²¹⁾	**5** ⁽²⁾	**3**
1	**2**	**3**	**5**	**4**	**6**

3 ⁽²⁾	**6** ⁽¹²⁰⁾	**5**	**4**	**2** ⁽⁶⁾	**1**
6	**4** ⁽³²⁾	**2** ⁽⁶⁰⁾	**5**	**1** ⁽¹⁾	**3**
4	**2**	**6**	**1** ⁽³⁾	**3** ⁽⁸⁾	**5** ⁽¹²⁰⁾
1 ⁽³⁰⁾	**3** ⁽⁶⁰⁾	**4**	**2**	**5**	**6**
2	**5**	**1** ⁽¹⁰⁸⁾	**3**	**6**	**4**
5	**1**	**3**	**6**	**4** ⁽²⁾	**2**

4 ⁽⁴⁾	**2** ⁽¹⁴⁾	**5**	**3**	**6** ⁽¹¹⁾	**1** ⁽¹⁸⁾
3 ⁽⁴⁵⁰⁾	**1** ⁽³⁾	**2**	**4**	**5**	**6**
5	**4** ⁽⁴⁾	**1** ⁽⁵⁾	**6**	**2** ⁽³⁰⁾	**3**
6	**5**	**4** ⁽¹²⁾	**1**	**3**	**2** ⁽²⁾
2 ⁽³⁾	**6**	**3**	**5**	**1** ⁽⁸⁰⁾	**4**
1 ⁽¹⁸⁾	**3**	**6**	**2** ⁽²⁾	**4**	**5**

1 ⁽⁵⁾	**6** ⁽¹²⁰⁾	**2** ⁽⁶⁾	**3**	**4** ⁽⁴⁾	**5** ⁽²⁾
6	**4**	**5**	**1**	**2** ⁽⁵⁾	**3**
3 ⁽¹⁵⁾	**5**	**6** ⁽¹⁴⁴⁾	**4**	**1**	**2**
5 ⁽⁸⁾	**2**	**4** ⁽⁴⁾	**6**	**3** ⁽³⁾	**1**
2 ⁽²⁾	**1**	**3** ⁽⁶⁾	**5** ⁽¹⁵⁰⁾	**6**	**4** ⁽²⁾
4	**3** ⁽³⁾	**1**	**2**	**5**	**6**

4 ⁽⁷²⁾	**3**	**6**	**1** ⁽⁵⁾	**2** ⁽²⁾	**5** ⁽⁴⁰⁾
1 ⁽¹⁾	**5** ⁽⁹⁾	**3**	**4**	**6** ⁽¹⁸⁰⁾	**2**
3 ⁽⁶⁾	**2**	**1**	**6**	**5**	**4**
2 ⁽⁶⁰⁾	**1** ⁽⁵⁾	**4** ⁽⁴⁾	**5** ⁽⁶⁰⁾	**3** ⁽²⁾	**6**
5	**6**	**2** ⁽⁷⁾	**3**	**4**	**1** ⁽¹⁾
6	**4** ⁽⁴⁾	**5**	**2** ⁽⁶⁾	**1**	**3**

1 ⁽⁹⁶⁾	**4**	**2**	**3** ⁽⁴⁵⁾	**5**	**6** ⁽¹¹⁾
4	**2** ⁽²⁾	**6** ⁽⁵⁾	**1**	**3**	**5**
3	**6** ⁽⁶⁾	**1** ⁽²⁵⁾	**5**	**4** ⁽²⁾	**2**
2 ⁽¹⁰⁾	**1**	**5**	**4** ⁽¹⁶⁾	**6**	**3** ⁽²⁴⁾
5	**3**	**4** ⁽¹²⁾	**6**	**2**	**1**
6 ⁽¹¹⁾	**5**	**3**	**2** ⁽³⁾	**1**	**4**

73, 74, 75, 76, 77, 78, 79, 80

81

180 5	10 4	6	2 2	1	3 3
2	40 5	4	1 1	3 3	6
3	2	120 5	6	4	40 1
6	6 1	2	3	5 5	4
5 1	6	3 3	2 4	2	5
4 4	3 3	1	1 5	6	2

82

180 5	6	2 4	2	3 3	1
6	15 3	5 1	1 4	5	2 2
2 2	5	6	3 3	7 1	1 4
4	1	150 5	6	2	3
1 1	5 2	3	5	4	180 6
7 3	4	3 2	1	6	5

83

288 3	4	6	1 1	10 5	2
1	6 6	4 2	12 3	4	5
4	2	1	100 5	2 6	3
3 2	1	5	4	11 3	5 6
30 5	8 3	7 4	6	2	1
6	5	3	2 2	1	4 4

84

1 1	16 2	5	72 3	4	6
160 4	3 3	1	6	2	1 5
2	1	14 3	5	6 6	4
14 3	5	6	7 4	2 1	2
6	4	48 2	1	45 5	3
5	6	4	2	3	1

85

9 3	12 2	1	5	24 4	6
2	4	90 3	1	11 6	5
12 1	5	6	3	40 2	5 4
6	3	2	4	5	1
9 5	5 6	4 4	2 2	1	8 3
4	1	1 5	6	3	2

86

8 3	4	1 1	1800 2	5	6
1	7 3	4	6	3 2	20 5
2 2	14 6	3	5	1	4
3 6	2	5	72 1	7 4	5 3
200 5	5 1	6	4	3	2
4	5	2	3	6	1

87

10 5	2	3 3	240 1	4	6
6 1	3	2	15 4	90 6	5
1 3	24 4	5	6	1	2
4	6	2 1	2	5	72 3
1 2	1	10 6	5 5	3	4
30 6	5	4	3	2	1

88

72 4	13 6	5	2	3 3	1
6	3	1 2	5 5	1	96 4
10 2	5	3	1	4	6
6 1	2	24 4	1 3	1 6	5
3	1	6	4	30 5	2
9 5	4	5 1	6	2 2	3

89

6 [54]	3	1 [14]	5 [30]	4 [9]	2 [2]
3	2 [2]	4	6	5	1
2	4	3	1 [5]	6	5 [1]
4 [9]	5	2 [10]	3 [6]	1	6
1 [5]	6	5	2	3 [24]	4
5 [4]	1	6 [2]	4	2	3 [3]

90

1 [5]	3 [60]	4	5	6 [17]	2 [2]
4	1 [2]	2	3 [36]	5	6
3 [2]	4 [1]	5	6	2	1 [8]
5	2 [11]	6	1 [1]	3	4
2 [19]	6	3 [3]	4	1	5 [15]
6	5	1 [7]	2	4	3

91

6 [450]	5	1 [2]	2 [19]	3	4
5	3	2	4	1 [30]	6
1 [5]	6	4 [4]	3 [45]	5	2 [8]
4 [12]	2	3	5	6	1
3	4 [11]	6	1	2 [2]	5
2	1	5 [15]	6	4	3 [3]

92

6 [14]	4	3 [15]	5	1 [2]	2 [2]
4	5 [21]	6	1 [14]	2	3
1 [6]	6	4	2	3	5
5	3 [30]	2	6 [6]	4 [13]	1 [5]
3 [7]	2	1	4	5	6
2	1	5	3 [72]	6	4

93

2 [2]	4 [3]	5 [90]	1 [18]	3	6
4	1	3	6	5 [1920]	2
3 [2]	5 [40]	2	4	6	1 [4]
6	2 [6]	1	3	4	5
1 [4]	3 [2]	6 [120]	5	2	4
5	6	4	2 [2]	1	3 [3]

94

4 [2]	6 [5]	1	3 [6]	2	5 [20]
2	5 [15]	3	1	6 [13]	4
3 [3]	1 [6]	4 [24]	6	5	2
1	2	5 [15]	4	3 [72]	6
5 [5]	3	6	2 [8]	4	1 [2]
6 [48]	4	2	5	1	3

95

2 [13]	4	1 [3]	3	6 [1]	5 [15]
3	2 [4]	6	1	5	4
1	3	2 [5]	5 [1]	4 [2]	6
4 [3600]	5	3	6	2	1 [6]
6	1 [6]	5	4 [32]	3 [3]	2
5	6	4	2	1	3

96

5 [15]	2 [2]	6 [2]	3	1 [7]	4
3	1	5 [15]	4	6	2
2 [2]	4	3 [8]	6 [17]	5	1
1 [1]	3	2	5	4 [2]	6
4 [120]	6 [9]	1	2	3 [45]	5
6	5	4 [5]	1	2 [2]	3

97

144 6	4	2	3	12 5	1
3 5	2	3 3	2 1	60 4	6
3 1	21 6	4	2	3	5
3	5	6	15 4	5 1	2
3 4	1	5	6	2	72 3
11 2	3	1	5	6	4

98

2 2	864 1	3	30 5	24 4	6
1	6	2	3	150 5	5 4
4	2	12 5	6	3	1
6	45 3	1	7 4	2	5
3	5	4 4	1	3 6	2
15 5	4	6	2	2 1	3

99

100 4	5	5 6	1	13 3	2 2
5	2 4	2	3	6	8 1
14 2	6	90 3	5	1	4
6	2	1	24 4	3 5	3
1	3	20 4	6	2	1 5
3 3	1	5	2 2	4	6

100

2 2	60 1	5	15 3	4	6
3	4	36 6	10 5	1	2
1	2	3	4	1 6	5
20 5	144 6	4	2	8 3	1
6	3	4 2	1	5 5	4
4	5	1	3 6	2	3 3

101

30 6	5	15 1	2 2	3	4
90 5	3	2	4	1	12 6
1	6	1 4	2 3	5	2
9 4	2	3	11 5	6	1
3	4 4	48 6	1	2	8 5
2 2	1	1 5	6	4	3

102

5 5	1	3 2	6	72 3	4
1	22 2	4	5	6	30 3
3 6	5	1 1	7 3	4	2
3	6	11 5	2 4	1 2	1
24 4	3	6	2	1	5
2	13 4	3	1	5	6 6

103

6 6	40 1	4	5	2	2 3
12 1	15 6	2 2	4	90 3	5
2	4	1 1	54 3	5	6
4	5	3	6	3 1	2 2
5	2 3	6	12 2	4	1
10 3	2	5	1	6	4 4

104

6 2	3	1	10 4	5	4 6
3 3	300 6	5	1	11 4	2
2 4	5	2	6	1	3 3
6	2880 2	4	5	3	1
30 1	4	3 3	7 2	6	1 5
5	1	6	3	2	4

105

[11] 4	1	3	[15] 6	[50] 2	5
2	[9] 3	1	4	5	[5] 6
[3] 3	4	2	5	[2] 6	1
[4320] 6	[12] 2	5	[6] 1	3	[1] 4
1	5	6	2	[4] 4	3
5	6	4	3	[2] 1	2

106

[26] 6	5	3	[3] 4	1	[2] 2
4	[4] 2	6	[2] 5	3	1
5	[2] 1	2	[6] 3	[1] 4	[2] 6
3	[24] 6	1	2	5	4
[2] 2	4	[20] 5	1	[14] 6	[2] 3
1	[3] 3	4	6	2	5

107

[100] 1	5	[2] 6	[2] 4	[6] 3	2
5	4	3	2	1	[5] 6
[2] 3	[120] 6	2	5	[4] 4	1
6	2	[1350] 1	3	5	[4] 4
[9] 2	3	4	[12] 1	6	5
[4] 4	[4] 1	5	6	2	3

108

[480] 4	5	[5] 1	[1] 3	2	[15] 6
[10] 2	3	5	1	[10] 6	4
1	2	[2] 3	[4] 6	4	5
5	4	6	2	[3] 3	1
[108] 6	1	[2] 2	[60] 4	5	3
3	6	4	[6] 5	1	[2] 2

109

[180] 5	1	[13] 6	[24] 2	3	4
2	3	[6] 4	6	[4] 1	[1] 5
1	[3] 2	[60] 3	4	5	6
3	6	[9] 1	5	[8] 4	2
6	[4] 4	5	3	2	[18] 1
[4] 4	5	2	1	6	3

110

[48] 2	6	4	[9] 1	5	3
[3] 3	[1] 1	[3] 6	[20] 4	[144] 2	[1] 5
1	[18] 2	3	5	4	6
[11] 6	4	[15] 5	3	1	[2] 2
5	3	[1] 1	[4] 2	6	4
4	5	2	6	3	1

111

[120] 5	4	[5] 6	[1] 1	2	3
6	[17] 3	5	2	1	4
[18] 1	6	[4] 3	[1] 4	5	2
[2] 3	5	[3] 2	6	4	[15] 1
[8] 4	[2] 2	1	5	3	6
2	1	[7] 4	3	[11] 6	5

112

[80] 5	4	[5] 1	6	[1] 2	3
4	[13] 6	5	[17] 2	3	1
[2] 1	2	[72] 3	4	[5] 6	5
2	[16] 3	6	[75] 5	1	4
[2] 6	1	[4] 4	3	5	2
3	5	2	1	4	[6] 6

113

600: 6	2	5	8: 3	1	2: 4
5	7: 1	2	4	12960: 3	6
2	4	5: 6	1	5	3: 3
3: 4	5: 5	5: 3	2	6	8: 1
1	3	4	6	2	5
2: 3	6	1: 1	1: 5	4	2

114

480: 2	4	5	5: 1	2: 6	3
4	1: 2	1	6	1800: 3	7: 5
3	108: 6	160: 4	5	1	2
6	3	2	4	5	48: 1
6: 1	5	3	2	4	6
5	18: 1	6	3	2	4

115

576: 3	6	4	2	6: 5	1
1	6: 5	14: 6	4	5: 2	3
2: 2	1	3	5	1440: 6	4
4	1: 3	2	5: 6	1	5
6: 6	20: 4	5	3: 1	3	2
7: 5	2	3: 1	3	4: 4	6

116

90: 5	3	3: 2	2: 6	4	3: 1
6	2: 2	5	8: 4	1	3
2: 2	4	17: 6	1	3	9000: 5
2: 3	6	1	2: 2	5	4
1	10: 5	4	3	2	14: 6
4	1	3	5	6	2

117

3: 1	25: 6	4	5: 2	3	11: 5
3	1	6	5: 5	4	2
1: 2	3	1	48: 4	11: 5	6
15: 4	5	3	6	2	8: 1
6	2: 2	15: 5	3	5: 1	4
5	4	2: 2	1	6	3

118

20: 5	1	4: 2	2: 3	6	4: 4
4	5: 5	6	12: 2	3	1
9: 1	2	5: 5	6	24: 4	3
6	16: 4	3	10: 5	1	2
1: 3	6	16: 4	1	2	150: 5
2	3	1	4	5	6

119

1: 3	72: 2	6	9: 5	1	1: 4
4	6	8: 1	2	3	5
10: 2	3: 3	2: 5	4	864: 6	2: 1
5	6: 1	3	6	4	2
1	4	40: 2	3	14: 5	6
6: 6	5	4	1	2	3

120

50: 5	2	1	2: 4	6	7: 3
1: 2	5	1: 4	3: 6	3	1
3	6: 6	5	2	30: 1	24: 4
3: 4	1	3: 3	5	2	6
1	480: 4	6	3	3: 5	2
6: 6	3	2	1	4	5

133

121

7 (22+)	5	6 (3−)	3	1 (320×)	8
3	7	1 (3+)	8	5	6 (1−)
1 (5−)	6 (6)	3	5 (280×)	8 (24×)	7
6	1 (120×)	8	7	3	5 (5)
8	3	5 (30×)	6	7 (7)	1 (18×)
5	8 (15+)	7	1	6	3

122

8 (3−)	6 (144×)	1 (8+)	7	5 (5)	3 (3+)
5	8	7 (42×)	3 (2+)	6	1
3	1	6	8 (1−)	7	5 (30×)
7 (735×)	3	5 (3−)	1 (48×)	8	6
1 (6×)	5	8	6	3 (3+)	7 (1−)
6	7	3 (2−)	5	1	8

123

1 (5−)	7 (42×)	6	8 (5−)	3	5 (280×)
6	1 (3+)	3	5 (13+)	8	7
5 (450×)	6	7 (21×)	3	1 (27+)	8
8 (5−)	3	1	7 (7)	5	6 (108×)
3	5	8	6	7	1
7 (280×)	8	5	1	6	3

124

5 (5)	6 (5−)	8 (40×)	1 (2−)	3	7 (1−)
3 (2÷)	1	5	6 (13+)	7	8
6	5 (2−)	3	7 (56×)	8 (9+)	1
7 (16+)	3 (3+)	1	8	5 (11+)	6
1	8 (56×)	7 (42×)	3 (2+)	6	5 (75×)
8	7	6	5	1	3

125

9 (7−)	6 (3−)	4 (512×)	2	8	7 (3−)
2	9	8	6 (15+)	7 (9+)	4
6 (2−)	4 (9408×)	7	9	2	8 (21+)
8	2 (11+)	6	7	4	9
4 (224×)	7	2	8	9 (54×)	6
7	8	9 (5−)	4	6 (3÷)	2

126

2 (5−)	8 (1−)	6 (42×)	7	9 (15+)	4 (2+)
7	9	4 (2+)	8	6	2
6 (2−)	7 (5−)	9 (5−)	4	2 (4÷)	8 (1−)
4	2	7 (5−)	6 (3−)	8	9
8 (4−)	4	2	9	7 (336×)	6
9 (23+)	6	8	2	4	7 (7)

127

8 (48×)	6	4 (2+)	2	9 (2−)	7
7 (11+)	9 (1728×)	6	8 (512×)	4	2 (3÷)
4	8	9 (2−)	7	2	6
2 (108×)	4	7 (42×)	6	8	9 (288×)
6	2 (5−)	8 (19+)	9	7 (42×)	4
9	7	2	4 (4)	6	8

128

8 (112×)	2	7	9 (6804×)	6	4 (4)
9 (17+)	8	6 (48×)	4 (13+)	7	2
4 (2−)	6 (108×)	8	7	2	9
6	9	2	8 (2+)	4	7 (21+)
2 (20+)	7	4 (2−)	6	9 (1−)	8
7	4	9 (7−)	2	8	6

129

3240× 9	15+ 7	8	567× 3	1− 5	4
8	9	5	7	3− 4	5− 3
20+ 4	5	3	9	7	8
7	4	22+ 9	8	3÷ 3	35× 5
480× 3	8	4	5	9	7
5	84× 3	7	4	1− 8	9

130

18+ 8	7	14+ 5	9	12× 4	3
9+ 5	3	1− 4	15+ 7	17+ 9	8
4	2520× 9	3	8	120× 5	7 7
3÷ 9	5	7	3	8	13+ 4
3	2+ 4	8	525× 5	7	9
7 7	8	13+ 9	4	3	5

131

19+ 4	3÷ 3	9	15+ 8	7	40× 5
7	5	108× 3	9	4	8
3	2+ 4	3− 8	5	4− 9	23+ 7
17+ 9	8	3− 4	7	5	3
8	7 7	12+ 5	4	3	9
315× 5	9	7	5− 3	8	4

132

35× 7	5	3÷ 9	3	20+ 4	8
2− 3	16+ 4	5	7	8	3÷ 9
5	17+ 7	4 4	22+ 8	9	3
2÷ 8	3	7	9 9	5	3− 4
4	648× 9	120× 8	5	3	7
9	8	1− 3	4	35× 7	5

133

18× 9	3− 2	5	27+ 3	8	6 6
2	2− 5	3	1− 9	6	8
1− 5	3− 6	9	8	3 3	2
6	19+ 9	8	3÷ 2	45× 5	2− 3
14+ 3	8	2	6	9	5
8 8	3	11+ 6	5	7− 2	9

134

3÷ 9	11+ 6	48× 8	5− 3	2700× 2	5
3	5	6	8	9	4÷ 2
6− 2	3÷ 9	1− 3	5	6	8
8	3	2	28+ 9	5	54× 6
30× 6	4+ 8	4− 5	3÷ 2	3	9
5	2	9	6	8	3

135

810× 9	6	3	5	720× 2	8
28+ 2	8	6	9	3 3	5
3	54× 2	5 5	2÷ 6	36+ 8	9
13+ 5	9	19+ 8	3	6	36× 2
8	3	9	2	5	6
6 6	3− 5	2	8	9	3

136

360× 8	17+ 2	6	9	11+ 3	1− 5
5	9	810× 2	3	8	6
3	5	9	48× 8	6	6− 2
3÷ 2	8+ 3	5	3− 6	9	8
6	24× 8	3	2 2	7+ 5	3÷ 9
9 9	19+ 6	8	5	2	3

137

1944× 9	6	4	2÷ 1	2	7 7
392× 7	4	9	13+ 2	1296× 1	6
1	2 2	5− 6	4	7	9
2	7	1	18+ 9	6	4
10+ 6	8− 9	2	7	5− 4	3+ 1
4	1	13+ 7	6	9	2

138

9 9	36× 4	13+ 6	7	14× 2	12+ 1
2÷ 2	9	5− 1	6	7	4
4	112× 2	8− 9	1	3888× 6	7
5− 6	1	7	4	9	3+ 2
1	20+ 7	2	9	36× 4	6
7	6	4	2	1	9

139

5− 1	2÷ 2	28× 7	4	19+ 6	9
6	1	20+ 9	14× 2	7	4
4 4	29+ 6	2	9	1	1− 7
5292× 2	4	1 1	16+ 7	9	6
7	9	4	6	9+ 2	1
9	7	6	1 1	4	2

140

7− 2	3+ 6	4 4	378× 1	20+ 9	7
9	2	1	6	7 7	4
5− 1	7	9	2 2	1728× 4	6
6	18+ 1	7	4	2	9 9
252× 7	4	6	9	2÷ 1	2
4	9	14× 2	7	7+ 6	1

141

32+ 6	8	7	5	180× 4	9
4 4	315× 7	3− 9	6	5	2÷ 8
5	9	6	56× 8	7	4
1− 8	11+ 6	5	3− 4	63× 9	7
9	5 5	8640× 4	7	48× 8	6
3− 7	4	8	9	6	5

142

22680× 7	15+ 8	30× 5	15+ 9	6	120× 4
9	7	6	2÷ 8	4	5
5	9	8	3− 4	7	6
30+ 4	5	7	2− 6	8	2520× 9
6	2− 4	36× 9	7 7	5	8
8	6	4	4− 5	9	7

143

12+ 7	5	168× 6	21+ 4	1− 8	9
2÷ 8	4	7	9	10080× 6	5
120× 5	15+ 7	4	8	5− 9	6
6	8	405× 9	18+ 5	4	7
4	9	5	6	7	8
432× 9	6	8	140× 7	5	4

144

5− 9	4 4	56× 7	8	6	1− 5
4	2− 6	8	29+ 5	2− 7	9
5	9	4	6	1− 8	7
15+ 7	8	24+ 9	4	5	6 6
3− 8	5	6	63× 7	9	2÷ 4
1− 6	7	45× 5	9	4 4	8

145

35×		16+		12+	5−
5	7	2	6	3	1
3÷					
3	1	5	2	7	6
3−		5−			15+
2	5	3	1	6	7
5−	2÷		35×		
6	2	1	7	5	3
	6	42×			
1	6	7	3	2	5
7	17+				
7	3	6	5	1	2

146

4−		4−		5−	105×
1	5	6	2	7	3
360×				5−	
3	6	2	7	1	5
	9+				6−
2	1	3	5	6	7
		35×	2÷		
5	2	7	6	3	1
4−			2÷		3÷
7	3	5	1	2	6
6	8+		8+		
6	7	1	3	5	2

147

84×			15×		
6	2	7	3	1	5
19+		4−		10×	
1	6	3	7	5	2
	6+		2÷		42×
2	1	5	6	3	7
		10×			
3	7	2	5	6	1
17+		5−		18+	
7	5	6	1	2	3
	6+				
5	3	1	2	7	6

148

11+		14×		36×	
5	1	2	7	6	3
10+		3÷			13+
7	5	1	3	2	6
		15×	30×		
1	2	3	6	5	7
1−			350×	3÷	
6	7	5	2	3	1
3−		18+		1−	
3	6	7	5	1	2
2	3	6	1	7	5

149

180×	17+			9	1152×
4	5	8	2	9	6
		17+			
9	2	5	8	6	4
	2−		5−		1−
5	6	2	9	4	8
4÷			3−		
8	4	6	5	2	9
	12960×		6	15+	
2	9	4	6	8	5
6					
6	8	9	4	5	2

150

5−		2÷	8	34+	
9	4	2	8	5	6
4÷	5		180×		11+
8	5	4	9	6	2
	4320×				
2	6	5	4	8	9
24×		23+			1−
4	2	8	6	9	5
			9+		
6	8	9	5	2	4
		6		4−	
5	9	6	2	4	8

151

2÷		432×			29+
8	4	2	9	6	5
11+					
5	6	4	8	2	9
18×	1−			10+	
2	8	9	5	4	6
	30×		1280×	4÷	
9	5	6	4	8	2
15+		8			
6	9	8	2	5	4
2−		1−		9	
4	2	5	6	9	8

152

288×		2÷		1−	
8	9	2	4	6	5
3−		5	72×		16+
2	4	5	8	9	6
	19+				
5	6	4	9	8	2
48×		19+			72×
4	2	8	6	5	9
	22+		2	32×	
6	5	9	2	4	8
9		1−			
9	8	6	5	2	4

153

7×	216×		1−	15+	21+
1	3	8	4	7	9
7	1	9	3	8	4
18144×			33+	1−	
4	7	3	9	1	8
3	36×	1	8	9	7
	4	196×		9×	
8	9	4	7	3	1
	8		4÷		
9	8	7	1	4	3

154

3×		24+			4
1	3	9	8	7	4
168×		96×	5−	54432×	
3	1	4	9	8	7
7	8	3	4	9	3÷
					1
8	18+	1	6−	4	3
	9		7		
1008×	7	8	1	3	9
4		19+			
9	4	7	3	1	8

155

21×	4÷	1−		16+	21+
3	1	9	8	7	4
1	4	4−	3	9	8
		7			
7	24×	3	4÷	4	9
	8		1		
24+	3	4−	4	21×	7
9		8		1	
4	37+	1	9	8	3
	7				
8	9	11+	7	3	1
		4			1

156

27×	168×			4÷	
9	7	3	8	1	4
3	1	20+	7	9	7−
		4			8
1−	3	19+	216×	8	1
4		7	9		
8+	4	8	3	16+	9
1				7	
7	18+	9	1	84×	3
	8			4	
1−	9	8+	4	3	7
8		1			

157

2100×					3
2	1	7	6	5	3
5	11+	1	16+	2	27+
	3	1	7		6
8+	2	2+	3	7	5
1		6			
7	6	3−	5	3	1
		2		3	
630×	5	6+	2	1	7
6		3			
3	7	4−	1	6	2
		5			

158

12×		8+		5	630×
6	2	1	7	5	3
1	17+	2	5	7	6
	3				
2−	5	7	1	3÷	2
3				6	
5	882×	3	6	2+	1
	7		6	2	
2	6	1−	6×	1	15+
2		5	3		7
7	1	6	2	3	5

159

11+	14×	15×	1−		3+
5	2	3	6	7	1
6	7	5	14×	6+	3
			2	1	
5−	2−	5−	7	5	4−
2	3	1			6
7	5	6	2−	1−	2
			1	3	
3÷	5−	5−	3	2	2−
1	6	7			5
3	1	2	1−	6	7
			5		

160

9+	6×	3÷	10×		2−
7	6	1	5	2	3
2	1	3	7+	2−	5
			6	7	
6×	2	15+	1	5	6
3		7			6
1	3	5	84×	6	7
1			2		
28+	1−	6	3	1−	2
5	7			1	
6	5	2	7	2−	1
				3	

161

192× 4	6	3− 2	5	34020× 9	7
210× 7	2	4	6	5	252× 9
6	5	22+ 9	2	7	4
2÷ 2	4	7	9	1− 6	5
405× 5	9	6	7 7	12+ 4	2
9	12+ 7	5	2− 4	2	6

162

5− 9	3− 4	10080× 5	7	6	2 2
4	7	6	7− 9	2	270× 5
21+ 5	9	7 7	2	4	6
2	5	36× 4	2− 6	3150× 7	9
1− 6	3÷ 2	9	4	5	28× 7
7	6	2	5	9	4

163

63× 9	5− 7	2	1− 4	5	3− 6
7	22+ 5	4	33+ 6	2	9
4− 2	9	150× 6	5	7	2÷ 4
6	4	5	7	9	2
20× 5	3÷ 6	567× 9	2	2− 4	12+ 7
4	2	7	9	6	5

164

60× 5	2+ 2	4	15+ 6	9	29+ 7
2	4	9	5 5	7	6
6	3− 9	2	5− 7	4	5
36× 9	6	21+ 7	2	1− 5	15+ 4
4	210× 7	5	9	6	2
7 7	5	6	2+ 4	2	9

165

26+ 4	8	1	162× 9	6	3
3	18+ 4	8	6	8− 1	9
1	9	5− 4	7+ 3	2− 8	6
7776× 8	6	9	4	3÷ 3	1
9	2÷ 3	6	4÷ 1	4	8 8
6	1	3	21+ 8	9	4

166

4÷ 4	21+ 9	8	1	12× 3	2− 6
1	3	72× 6	23+ 9	4	8
3	1	4	6	8	27× 9
144× 6	8	3	288× 4	9	1
1− 9	3888× 4	1 1	8	6	3
8	6	9	3	1	4 4

167

2− 6	8	4÷ 1	4	3÷ 3	9
10+ 1	6	34992× 9	3	5− 2÷ 4	8
3	3− 4	6	8	10+ 9	1
72× 8	1	4	9	6	3
9	72× 3	8	5− 6	1	2− 4
13+ 4	9	3	7− 1	8	6

168

12× 4	3	13+ 9	144× 8	6	1
17+ 9	8	4	8− 1	3	6 6
3− 3	10+ 4	6	9	18+ 1	8
6	29+ 1	3	4	8	9
7− 1	48× 6	8	2+ 3	9	1− 4
8	8− 9	1	6	4	3

169

35+ 6	9	3	1280× 8	4	5
9	2− 4	6	2− 5	3	8
3	5	192× 4	6	8	3÷ 9
69120× 4	6	22+ 8	9	5	3
8	15× 3	5	36× 4	9	16+ 6
5	8	9	3 3	6	4

170

720× 9	2− 5	3	14+ 8	6	1− 4
5	4	26+ 6	9	8	3
4	16200× 6	5	3	9	72× 8
1296× 6	8	4	5	3	9
3	9	2+ 8	4	150× 5	6
11+ 8	3	9 9	24× 6	4	5

171

31104× 4	4− 5	2+ 3	6	21+ 9	8
3	9	6	160× 8	5 5	4
6	8	9	5	4	3− 3
40× 5	12+ 3	4	12960× 9	8	6
8	4 4	5	3	15+ 6	9
9 9	2− 6	8	4	3	5

172

90× 3	5	6 6	1− 8	9	13+ 4
6	3 3	192× 4	3− 5	8	9
5 5	6	8	21+ 9	1− 4	33+ 3
2÷ 8	36× 4	9	3	5	6
4	9	30× 5	6	3	8
17+ 9	8	3	4	6	5 5

173

5 5	18+ 9	2	280× 7	8	3+ 3
96× 8	2	7	5	3÷ 3	9
2	3	41+ 5	80× 8	9	7 7
3+ 3	7	9	2	5	4÷ 8
9	5	5− 8	6− 3	14+ 7	2
7	8	3	9	2	5

174

5− 3	8	9450× 2	5	9	7 7
7	3	5	13+ 2	8	36+ 9
16+ 5	2	147× 7	9 9	3	8
4+ 8	9	3	7	2	5
2	5 5	216× 9	8	7	6× 3
24+ 9	7	8	3	5	2

175

4÷ 2	88200× 9	5	7	30× 3	8 8
8	7	3 3	16+ 9	2	5
3+ 3	8	18+ 7	2	5	36+ 9
9	5	8	3	7 7	2
210× 7	2	9	5	8	3
5	3	4÷ 2	8	2− 9	7

176

2− 3	54× 2	9	20+ 5	7	4÷ 8
5	3	4− 7	19+ 9	8	2
56448× 7	9	3	8	2	2− 5
9 9	7	4+ 8	2	2835× 5	3
2	8	50× 5	63× 3	9	7
8	5	2	7	3	9

177

27+ 5	7	1	4	90× 9	2
1	5- 2	7	9	18+ 4	5
7- 2	4- 1	3- 4	7	5	5- 9
9	5	2÷ 2	1	7	4
3- 7	5- 4	9	450× 5	2	7× 1
4	9	5	2	1	7

178

98× 2	7	4- 5	1	20+ 4	9
7	2 2	4÷ 1	3- 4	4- 9	5
8- 1	18+ 9	4	7	5	2
9	5	9+ 7	2 2	8+ 1	2520× 4
20× 5	4	2	9	7	1
4	1	9 9	5	2	7

179

28× 1	16+ 4	7	5	2÷ 2	315× 9
4	7	7- 2	5- 9	1	5
3- 2	5	9	4	7	1
18× 9	2	1	1960× 7	5	4
26+ 7	9	5	1	5- 4	2
5	7+ 1	4	2	9	7

180

9 9	2÷ 1	2	33+ 4	5	7
14× 1	2	7	9	1- 4	5
504× 2	4	20× 5	1	7	8- 9
7	9	4	2520× 5	2	1
21+ 5	7	9	2	4÷ 1	4
10+ 4	5	1	7	9	2

Printed in the United States
By Bookmasters